Single Nucleotide Polymorphisms

METHODS IN MOLECULAR BIOLOGY™

John M. Walker, SERIES EDITOR

METHODS IN MOLECULAR BIOLOGY™

Single Nucleotide Polymorphisms

Methods and Protocols

Edited by

Pui-Yan Kwok, MD, PhD

*Cardiovascular Research Institute
and Department of Dermatology
University of California, San Francisco
San Francisco, CA*

Humana Press ✳ Totowa, New Jersey

Library of Congress Cataloging-in-Publication Data

Single nucleotide polymorphisms ; methods and protocols / edited by Pui-Yan Kwok.

　　p. cm. -- (Methods in molecular biology ; 212)
　　Includes bibliographical references and index.
　　ISBN 0-89603-968-4 (alk. paper)
　　1. Chromosome polymorphism--Laboratory manuals. 2. Human genetics--Variation--Laboratory manuals. 3. Genetic markers--Laboratory manuals. I. Kwok, Pui-Yan, 1956– II. Methods in molecular biology (Totowa, N.J.) ; v. 212

QH447.6.S565 2002
611'.01816--dc21

2002024055

Preface

With the near-completion of the human genome project, we are entering the exciting era in which one can begin to elucidate the relationship between DNA sequence variation and susceptibility to disease, as modified by environmental factors. Single nucleotide polymorphisms (SNPs) are by far the most prevalent of all DNA sequence variations. Although the vast majority of the SNPs are found in noncoding regions of the genome, and most of the SNPs found in coding regions do not change the gene products in deleterious ways, SNPs are thought to be the basis for much of the genetic variation found in humans. As explained eloquently by Lisa Brooks in Chapter 1 of *Single Nucleotide Polymorphisms: Methods and Protocols*, SNPs are the markers of choice in complex disease mapping and will be the focus of the next phase of the human genome project. Besides the obvious applications in human disease studies, SNPs are also extremely useful in genetic studies of all organisms, from model organisms to commercially important plants and animals.

Identification of SNPs has been a laborious undertaking. In *Single Nucleotide Polymorphisms: Methods and Protocols*, the inventors of the most successful mutation/SNP detection methods (including denaturing high-performance liquid chromatography [dHPLC], single-strand conformation polymorphism [SSCP], conformation-sensitive gel electrophoresis [CSGE], chemical cleavage, and direct sequencing) describe the most current protocols for these methods. In addition, a chapter on computational approaches to SNP discovery in sequence data found in public databases is also included.

Genotyping SNPs has been a particularly fruitful area of research, with many innovative methods developed over the last decade. The second half of *Single Nucleotide Polymorphisms: Methods and Protocols* contains chapters written by the inventors of the most robust SNP genotyping methods, including the molecular beacons, Taqman assay, single-base extension approaches, pyrosequencing, ligation, Invader assay, and primer extension with mass spectrometry detection. Since the projected need for SNP genotyping is in the order of 200 million genotypes per genome-wide association study, methods described in this volume will form the basis of ultrahigh-throughput genotyping approaches of the future.

I am indebted to a most talented group of friends and colleagues who have put together easy-to-follow protocols of the methods they invented for this volume. It is my hope that *Single Nucleotide Polymorphisms: Methods and Protocols* will serve as a guidebook to all interested in SNP discovery and genotyping and will inspire innovative minds to develop even more robust methods to make complex disease mapping and molecular diagnosis a reality in the near term.

Pui-Yan Kwok, MD, PhD

Contents

Contributors

JEFFREY J. BABON • *Genomic Disorders Research Centre, St. Vincent's Hospital, Melbourne, Victoria, Australia*

DIRK VAN DEN BOOM • *Sequenom Inc., San Diego, CA*

LISA D. BROOKS • *National Human Genome Research Institute, National Institutes of Health, Bethesda, MD*

CHINH T. BUI • *Genomic Disorders Research Centre, St. Vincent's Hospital, Melbourne, Victoria, Australia*

RICHARD G. H. COTTON • *Genomic Disorders Research Centre, St. Vincent's Hospital, Melbourne, Victoria, Australia*

BRIGITTE DARNHOFER-PATEL • *Sequenom Inc., San Diego, CA*

SHENGHUI DUAN • *Division of Dermatology, Washington University, St. Louis, MO*

ARUPA GANGULY • *Department of Genetics, University of Pennsylvania, Philadelphia, PA*

SØREN GERMER • *Roche Molecular Systems, Alameda, CA*

KENSHI HAYASHI • *Division of Genome Analysis, Research Center for Genetic Information, Medical Institute of Bioregulation, Kyushu University, Higashi-ku, Fukuoka, Japan*

RUSSELL HIGUCHI • *Roche Molecular Systems, Alameda, CA*

TONY M. HSU • *Division of Dermatology, Washington University, St. Louis, MO*

JONAS JARVIUS • *Rudbeck Laboratory, Unit of Molecular Medicine, Department of Genetics and Pathology, Uppsala University, Uppsala, Sweden*

FRED RUSSELL KRAMER • *Department of Molecular Genetics, Public Health Research Institute, Newark, NJ*

YOJI KUKITA • *Division of Genome Analysis, Research Center for Genetic Information, Medical Institute of Bioregulation, Kyushu University, Higashi-ku, Fukuoka, Japan*

PUI-YAN KWOK • *Cardiovascular Research Institute and Department of Dermatology, University of California, San Francisco, San Francisco, CA*

ANDREANA LAMBRINAKOS • *Genomic Disorders Research Centre, St. Vincent's Hospital, Melbourne, Victoria, Australia*

ULF LANDEGREN • *Rudbeck Laboratory, Unit of Molecular Medicine, Department of Genetics and Pathology, Uppsala University, Uppsala, Sweden*

ULRIKA LILJEDAHL • *Department of Medical Sciences, Uppsala University; Uppsala University Hospital, Uppsala, Sweden*

KATARINA LINDROOS • *Department of Medical Sciences, Uppsala University; Uppsala University Hospital, Uppsala, Sweden*

KENNETH J. LIVAK • *Applied Biosystems, Foster City, CA*

VICTOR LYAMICHEV • *Third Wave Technologies Inc., Madison, WI*

SALVATORE A. E. MARRAS • *Department of Molecular Genetics, Public Health Research Institute, Newark, NJ*

GABOR T. MARTH • *National Center for Biotechnology Information, National Library of Medicine, National Institutes of Health, Bethesda, MD*

BRUCE NERI • *Third Wave Technologies, Inc., Madison, WI*

MATS NILSSON • *Rudbeck Laboratory, Unit of Molecular Medicine, Department of Genetics and Pathology, Uppsala University, Uppsala, Sweden*

PETER J. OEFNER • *Stanford Genome Technology Center, Palo Alto, CA*

CHARLOTTA OLSSON • *Department of Medical Sciences, Uppsala University; Uppsala University Hospital, Uppsala, Sweden*

ANDREAS PREMSTALLER • *Stanford Genome Technology Center, Palo Alto, CA*

CHARLES P. RODI • *Rodi Pharma, San Diego, CA*

MOSTAFA RONAGHI • *Stanford Genome Technology Center, Palo Alto, CA*

NIELS STORM • *Sequenom GmbH, Hamburg, Germany*

AKARI SUZUKI • *Division of Genome Analysis, Research Center for Genetic Information, Medical Institute of Bioregulation, Kyushu University, Higashi-ku, Fukuoka, Japan*

ANN-CHRISTINE SYVÄNEN • *Department of Medical Sciences, Uppsala University; Uppsala University Hospital, Uppsala, Sweden*

TOMOKO TAHIRA • *Division of Genome Analysis, Research Center for Genetic Information, Medical Institute of Bioregulation, Kyushu University, Higashi-ku, Fukuoka, Japan*

SANJAY TYAGI • *Department of Molecular Genetics, Public Health Research Institute, Newark, NJ*

1

Snps: *Why Do We Care?*

Lisa D. Brooks

1. Introduction

Single-nucleotide polymorphism (SNP) is a new term for an old concept. Geneticists have been trying for decades to find the genetic differences among individuals. Originally phenotypes were used, then protein sequence, electrophoresis, restriction fragment polymorphisms (RFLPs), and microsatellites. With recent technologies for DNA sequencing and the detection of single-base differences, we are approaching the time when all differences in DNA sequence among individuals can be found. The next challenge is to relate these genetic differences to phenotypes such as disease risk and response to therapies.

2. Types of SNPs

SNPs most commonly refer to single-base differences in DNA among individuals. The assays that detect these point differences generally can also detect small insertions or deletions of one or a few bases. Polymorphisms are usually defined as sites where the less common variant has a frequency of at least 1% in the population, but for some purposes rarer variants are important as well.

From: *Methods in Molecular Biology, vol. 212:*
Single Nucleotide Polymorphisms: Methods and Protocols
Edited by: P-Y. Kwok © Humana Press Inc., Totowa, NJ

SNPs are useful for finding genes that contribute to disease, in two ways. Some SNP alleles are the actual DNA sequence variants that cause differences in gene function or regulation that directly contribute to disease processes. Most SNP alleles, however, probably contribute little to disease. They are useful as genetic markers that can be used to find the functional SNPs because of associations between the marker SNPs and the functional SNPs.

SNPs of various types can change the function or the regulation and expression of a protein. The most obvious type is a nonsynonymous SNP, where the alleles differ in the amino acid of the protein product. Some SNPs are polymorphisms at splice sites, and result in variant proteins that differ in the exons they contain *(1)*. Some SNPs are in promoter regions and are reported to affect the regulation and expression of proteins *(2–5)*. Caution is needed when trying to assign causality to a SNP as being the difference that directly affects protein function or expression. When SNPs are associated with other SNPs because of linkage disequilibrium, then many SNPs, in exons, introns, and other noncoding regions, may all be associated with a disease or phenotype, even though only one or a few may directly affect the phenotype.

3. Number of SNPs

How many SNPs are there in the human genome? This is the same as asking how many of the 3.2 billion sites in the genome have variant forms, at frequencies above the mutation rate.

There is good information on the proportion of sites that differ between two randomly chosen homologous chromosomes. This proportion is called the nucleotide diversity; it is useful for comparing the amount of variability among chromosome regions or among populations, and takes into account the number of chromosomes examined *(6)*. Many SNPs were discovered in the overlap of the ends of BAC clones used to assemble the human genome, when these BAC clones came from different individuals or from different chromosomes from the same individual; the number of differences between two chromosomes averaged 1/1331 sites of the DNA sequence *(7)*. Since people have two copies of all chromosomes (except the sex chromosomes in males), this means that any one individual is heterozygous at about 3.2 billion bases $\times$ 1 difference/1331 bases = 2.4 million sites across all chromosomes.

When two chromosomes are compared, they may have the same base at a DNA site even though that site is polymorphic in the population. The number of sites that vary in a population cannot be estimated simply by counting the number of sites that differ between two chromosomes. The number of sites seen to have variants will rise as more individuals are examined; the exact number will depend on the distribution of the frequencies of the SNP alleles, but many SNPs will be missed. For example, samples of 10 chromosomes have a 97% chance of including both SNP alleles when the minor allele frequency is at least 20% in the population, but only a 59% chance when the minor allele frequency is at least 1% *(8)*. Thus small samples are going to miss many SNPs with common alleles as well as most SNPs with rare alleles, and even samples that are larger are going to miss many SNPs with rare alleles.

Based on neutral theory and the observed rate of 1/1331 differences in two chromosomes, the estimate of the number of SNPs in humans with minor allele frequencies above 1% is 11 million *(8)*. However, this estimate misses SNPs that are rare overall but are more common in some populations. Currently there is too little information about the variation in rare allele frequencies among populations as well as about the deviations from the assumptions of the neutral model to make a good guess of the number of SNPs *(9)*. A rough guess is that there are about 10–30 million SNPs in the human genome, or one on average about every 100–300 bases. Eventually the number of SNPs will be found empirically, as many individuals are genotyped across the genome.

Genes are quite different in how much variation they contain, especially in the coding regions. Two large studies examined SNPs in small areas around genes, including exons, introns, and 5' and 3' UTRs *(10,11)*. The number of SNPs found per gene ranged from

0–50. Cargill et al. *(10)* looked at an average of 1851 bases for 106 genes in an average of 114 copies of each gene and found a rate of SNPs of 1/348 sites, with an average of 5 SNPs per gene; Halushka et al. *(11)* looked at an average of 2527 bases for 75 genes in 148 copies of each gene and found a rate of SNPs of 1/242 sites, with an average of 10 SNPs per gene. The difference in the average number of SNPs per gene can be explained by the second study's examining more bases in more individuals with more diversity, in what happened to be a more highly variable set of genes.

Averaging across the genes in these two studies, synonymous SNPs were more common than nonsynonymous SNPs. Only 38% of the nonsynonymous SNPs were seen compared with the number expected if the SNPs were neutral, which is evidence of selection against variants that change an amino acid. The average minor allele frequency for nonsynonymous SNPs was lower than for other classes of SNPs, which means that the sample sizes needed to find such SNPs will be larger than those based on average SNP allele frequencies. In noncoding regions, the rate of SNPs was lower than expected under the neutral polymorphism rate, showing some evidence of selection for conservation of the sequence of noncoding regions. This result may have occurred because the noncoding regions were next to the coding regions and included conserved regulatory regions.

When particular gene regions are looked at over longer stretches, there is often much variation: 21 SNPs and 1 indel formed 31 haplotypes in 5,491 bases of the APOE gene region in 144 chromosomes *(12,13)*, which is a variant every 250 bases; 74 SNPs and 4 indels formed 13 haplotypes in 24,070 bases of the ACE gene region in 22 chromosomes *(14)*, which is a variant every 309 bases; 79 SNPs and 9 indels formed 88 haplotypes in 9,734 bases of part of the LPL gene region in 142 chromosomes *(15,16)*, which is a variant every 111 bases.

4. The Pattern of Human SNP Variation

Humans arose about 100,000–200,000 years ago in Africa, and spread from there to the rest of the world *(17)*. The original population was polymorphic, and so populations around the world share most polymorphisms from our common ancestors. For example, all populations are variable at the gene for the ABO blood group. About 85–90% of human variation is within all populations *(18)*. Thus any two random people from one population are almost as different from each other as are any two random people from the world.

Mutations have arisen in populations since humans spread around the world, so some variation is mostly within particular populations. Variants that are rare are likely to have arisen recently, and are more likely than common variants to be found in some populations but not others *(14,15)*. Common variants are usually common in all populations. Only a small proportion of variants are common in one population and rare in another. Usually, a difference among populations is of the sort that a variant has a frequency of 20% in one population and 30% in another.

Figure 1 shows this pattern of human variation. The large overlap among the circles shows that all populations contain mostly the same variation. The small nonoverlap regions are still important for population differences in susceptibility to disease, but even then not all people in a population get any particular disease. Most differences in disease risk are among individuals regardless of population, rather than among populations.

5. Using SNPs to Find Genes Associated with Diseases

Common diseases such as cancer, stroke, heart disease, diabetes, and psychiatric disorders are influenced by many genes as well as by environmental factors. The goal of finding genes that affect a disease is to be able to understand the processes that produce the disease, with the hope of then figuring out therapeutic interventions that will prevent or cure the disease. Because populations share most genetic variants, the common diseases are expected to be influenced by variants that are common in all populations *(19–21)*.

Relating SNPs to complex diseases is going to be challenging. The most appropriate experimental design depends on the genetic basis for a disease, such as the number of genes affecting the disease,

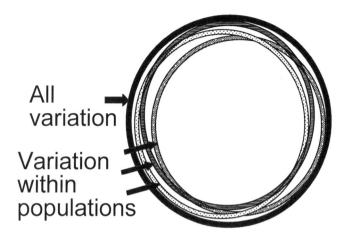

Fig. 1. Distribution of human variation within and between populations. The outer circle is the entire amount of human variation. Each
other circle shows the variation within one population.

the relative sizes of their contributions, the allele frequencies, and
the interactions between the genes; the amount of linkage disequilibrium around the genes; the types and amount of environmental influences; the interactions between the genetic and
environmental factors; and the genetic differences between control
and affected groups *(22)*. This information will be known better after
a study than before it. The genes and variants with the biggest effects
will be found most easily, and others should be found with the larger
sample sizes made possible by cheaper and more efficient technologies for genotyping.

SNPs with minor alleles of various frequencies are all useful. For
association analysis, researchers frequently want to use SNPs with
minor allele frequencies of at least 20%, so that the SNPs are informative about associations. However, common SNP alleles may
generally be old, so that recombination has had a longer time to
break down the associations around the SNPs *(23)*. The best power
in association studies comes when the marker SNP alleles and the
associated disease-contributing alleles are similar in frequency, so
including a range of SNP allele frequencies is useful. The SNP alle-

les that affect gene function, and so are generally selected against, will have lower average frequencies than alleles at other SNPs but may still be of interest as contributing to disease.

The technology is not yet cheap enough for studies that would genotype thousands of individuals for hundreds of thousands of SNPs across the genome in order to see which variants are most closely associated with a disease phenotype *(24)*. Looking at pooled samples, to find differences in frequency between affected and control groups, would reduce the number of samples per SNP to two, which would be a good screening tool to identify regions of the genome to analyze in more detail *(25)*. Currently researchers examine candidate genes they think are related to the disease process. This is an efficient method of examining likely suspects, but it misses genes with real but unknown contributions to the disease. Another cost-saving strategy is to focus on exons, but this risks missing regulatory variants.

Another method for increasing the efficiency of using SNPs is to determine haplotypes. Recent studies have shown that much of the genome is organized into blocks of haplotypes, with only a few haplotypes common in each region *(14,16,23,26,27)*. Just a few SNPs will suffice to mark these haplotype blocks and test whether they are associated with a disease. This block structure makes it easier initially to identify which chromosome regions are associated with the disease. However, once particular blocks are shown to be associated with the disease, then figuring out which genes and variants within the blocks are functionally causal becomes difficult because of the strong associations among SNPs within a block *(28)*. A large block may contain many genes; a smaller block may identify one gene, which is useful for understanding the disease process, but may still have many associated SNPs.

When multiple genes affect a disease, much more information is contained in haplotypes than in SNPs one at a time. Associations are better made with haplotypes than with single SNPs, because mutations occur on particular haplotype backgrounds and are associated with nearby SNPs until recombination or recurrent mutation breaks down these associations *(13)*. Even haplotypes do not contain the full information relating SNPs to diseases, because the diploid combination of haplotypes may also be important. An example occurs with type 2 diabetes; a pair of haplotypes contributes the highest risk jointly, although homozygotes for either haplotype have little increase in risk *(29)*.

Once small blocks of highly associated SNPs are identified as being associated with a disease, then statistical analysis is exhausted; it cannot identify which SNPs are functionally causal and which are statistically associated but not related to the disease. To identify the particular genes or SNPs functionally involved in the disease process requires either finding more samples with smaller blocks in the region *(30)*, or performing experiments. One type of experiment is to create SNP alleles in a constant background in a model organism *(31)*. Another type of experiment is to fill in the steps in the path from genotype to phenotype, by studying how different alleles cause functional differences, for example in gene expression patterns, protein amount and localization, protein structure or binding, or pathways. In contrast to the classical genetic approach of using knockouts to understand how genes work, using natural variants may provide more subtle information on how proteins function in health and disease.

6. Understanding the Distribution of SNPs

Understanding the distribution of SNPs will require understanding chromosome-level and population-level processes. The neutral theory of population genetics provides models generating the expected distributions of SNP allele frequencies and haplotype frequencies, given standard assumptions such as uniform mutation rates, specified population size or changes in size, and no selection *(6)*. These models are useful for comparing with observed data to figure out which assumptions are not true; which parameter values, such as population size, are most consistent with the data; and what types of selection may be occurring in particular chromosome regions.

A chromosome-level process that is important for SNP allele frequencies and linkage disequilibrium is the mutation rate, which is not uniform. Although SNPs in general have a low mutation rate, CpG dinucleotides are highly mutable; they form only about 1–2% of the sequence but about 25–30% of the SNPs *(11,32,33)*. Other types of mutation hotspots also exist, and gene conversion may also affect the frequencies of SNPs and the amount of linkage disequilibrium *(32)*. SNPs that arise by recurrent mutation may sometimes be at functionally important sites, and thus contribute to disease risk. However, SNPs that arise by recurrent mutation are going to be less informative as markers for association analyses because they are less associated with other SNPs.

Recombination is important for breaking down linkage disequilibrium. Haplotype blocks may reflect recombination hotspots, or simply historical recombination events. Regions with less recombination generally have lower amounts of genetic variation, as seen in humans, mice, and flies *(34–36)*. Presumably this reflects a history of selective sweeps for advantageous alleles or purifying selection against deleterious alleles, with low rates of recombination resulting in large regions of disequilibrium that get pulled along as natural selection changes the haplotype frequencies in chromosomal regions *(37)*.

SNPs can provide information on population history and on the form of selection on genes. The distribution of the number of mismatches between random individuals gives information about when population bottlenecks occurred in a population *(38)*. Comparing the ratio of synonymous to nonsynonymous changes in a gene within a species to the ratio of synonymous to nonsynonymous fixed differences between species provides information on the type of selection that has acted on genetic variation in the gene. Evidence for selection against variants in a gene occurs when there is an excess of synonymous fixed changes; evidence for balancing selection to keep variation in a population occurs when there is an excess of nonsynonymous fixed changes *(39)*. When variability is compared within and between two species, the expectation under the neutral model is that regions of high variability within both species correspond to regions of high divergence between species, reflecting simply a high mutation rate in those regions. Patterns inconsistent with this one may be evidence for natural selection of various sorts *(40)*. Demographic events, such as changes in population size, affect all genomic regions, while selective events affect particular genomic regions. Comparative population genomics, where the pattern of variability is compared between species, will provide insight into gene function and the processes that influence variation.

7. Methods That Will Be Needed

SNPs and other less common sequence variants are the ultimate basis for genetic differences among individuals, and thus the basis for most genetic contributions to disease. To make good use of SNPs for finding genes related to disease and studying their function, better and cheaper technological methods are needed for discovering SNPs, for genotyping them in many individuals, for finding their frequencies in pooled samples, and for discerning haplotypes. New statistical methods are needed to analyze linkage and association in large-scale studies, to relate haplotypes and the diploid genotypes they form to disease risk, and to elucidate the interactions among genes and between genes and the environment.

With the number of SNPs identified approaching 3 million, there will soon be enough to use as markers for linkage and association studies across the genome. The number of SNPs useable for these studies is smaller than the total number known, for several reasons: many SNPs have minor allele frequencies below the 20% most useful for linkage and association studies, the number of SNPs with minor allele frequencies above 20% in most populations is only about one-third of those with the minor allele above 20% in one population, some SNPs do not work well in assays, and SNPs that are near each other and highly associated do not provide independent information. Thus over the next couple of years the technologies for discovering new SNPs will still be important for finding

SNPs for linkage and association studies. Even if haplotypes turn out generally to have a block structure, the SNPs in the blocks are not going to be completely associated, so more SNPs than the number of common haplotypes in blocks will be needed to be reasonably sure of finding disease associations.

Even when a set of marker SNPs is found for linkage and association studies, the discovery of new SNPs will still be important. Researchers will want to know all the common and as many as possible of the less common SNPs in particular chromosome regions for studying the function of particular genes and for relating disease risk to variation in candidate genes or regions identified by whole genome scans. For these analyses, when researchers are interested in finding functional SNP variation, the rarer SNPs may be important, so methods of comprehensive SNP discovery in large samples will be needed.

For the best chance of associating gene regions and then particular genes and variants with diseases, large numbers of individuals are going to need to be genotyped for hundreds to thousands of SNPs. Cheap and efficient large-scale technologies for genotyping individual and pooled samples will allow the genetic contributions to be figured out even for the common diseases with complicated interactions of causes.

In the following chapters experts discuss the methods they developed to discover unknown SNPs and to genotype known SNPs. Different methods have different advantages and limitations, so the most appropriate method will vary depending on the particular experiment to be done. These chapters cover the range of current SNP discovery and genotyping methods for candidate gene regions and the entire genome, as the first step towards finding the genes contributing to disease and studying the disease process.

References

1. Krawczak, M., Reiss, J., and Cooper, D. N. (1992) The mutational spectrum of single base-pair substitutions in mRNA splice junctions of human genes: causes and consequences. *Hum. Genet.* **90**, 41–54.

2. El-Omar, E. M., Carrington, M., Chow, W.-H., McColl, K. E., Bream, J. H., Young, H. A., et al. (2000) Interleukin-1 polymorphisms associated with increased risk of gastric cancer. *Nature* **404**, 398–402.

3. Ligers, A., Teleshova, N., Masterman, T., Huang, W.-X., and Hillert, J. (2001) CTLA-4 gene expression is influenced by promoter and exon 1 polymorphisms. *Genes Immun.* **2**, 145–152.

4. Rutter, J. L., Mitchell, T. I., Butticé, G., Meyers, J., Gusella, J. F., Ozelius, L. J., and Brinckerhoff, C. E. (1998) A single nucleotide polymorphism in the matrix metalloproteinase-1 promoter creates an Ets binding site and augments transcription. *Cancer Res.* **58**, 5321–5325.

5. van der Pouw Kraan, T. C., van Veen, A., Boeije, L. C., van Tuyl, S. A., de Groot, E. R., Stapel, S. O., et al. (1999) An IL-13 promoter polymorphism associated with increased risk of allergic asthma. *Genes Immun.* **1**, 61–65.

6. Hartl, D. L. and Clark, A. G. (1997) *Principles of Population Genetics*, 3rd ed. Sinauer, Sunderland, MA.

7. The International SNP Map Working Group (2001) A map of human genome sequence variation containing 1.42 million single nucleotide polymorphisms. *Nature* **409**, 928–933.

8. Kruglyak, L. and Nickerson, D. A. (2001) Variation is the spice of life. *Nat. Genet.* **27**, 234–236.

9. Przeworski, M., Hudson, R. R., and Di Rienzo, A. (2000) Adjusting the focus on human variation. *Trends Genet.* **16**, 296–302.

10. Cargill, M., Altshuler, D., Ireland, J., Sklar, P., Ardlie, K., Patil, N., et al. (1999) Characterization of single-nucleotide polymorphisms in coding regions of human genes. *Nat. Genet.* **22**, 231–238.

11. Halushka, M. K., Fan, J.-B., Bentley, K., Hsie, L., Shen, N., Weder, A., et al. (1999) Patterns of single-nucleotide polymorphisms in candidate genes for blood-pressure homeostasis. *Nat. Genet.* **22**, 239–247.

12. Nickerson, D. A., Taylor, S. L., Fullerton, S. M., Weiss, K. M., Clark, A. G., Stengård, J. H., et al. (2000) Sequence diversity and large-scale typing of SNPs in the human apolipoprotein E gene. *Genome Res.* **10**, 1532–1545.

13. Fullerton, S. M., Clark, A. G., Weiss, K. M., Nickerson, D. A., Taylor, S.L., Stengård, J. H., et al. (2000) Apolipoprotein E variation at the sequence haplotype level: implications for the origin and maintenance of a major human polymorphism. *Am. J. Hum. Genet.* **67**, 881–900.

14. Rieder, M. J., Taylor, S. L., Clark, A. G., and Nickerson, D. A. (1999) Sequence variation in the human angiotensin converting enzyme. *Nat. Genet.* **22**, 59–62.

15. Nickerson, D. A., Taylor, S. L., Weiss, K. M., Clark, A. G., Hutchinson, R. G., Stengård, J., et al. (1998) DNA sequence diversity in a 9.7-kb region of the human lipoprotein lipase gene. *Nat. Genet.* **19**, 233–240.

16. Clark, A. G., Weiss, K. M., Nickerson, D. A., Taylor, S. L., Buchanan, A., Stengård, J., et al. (1998) Haplotype structure and population genetic inferences from nucleotide-sequence variation in human lipoprotein lipase. *Am. J. Hum. Genet.* **63**, 595–612.

17. Tishkoff, S. A., Dietzsch, E., Speed, W., Pakstis, A. J., Kidd, J. R., Cheung, K., et al. (1996) Global patterns of linkage disequilibrium at the CD4 locus and modern human origins. *Science* **271**, 1380–1387.

18. Barbujani, G., Magagni, A., Minch, E., and Cavalli-Sforza, L. L. (1997) An apportionment of human DNA diversity. *Proc. Natl. Acad. Sci. USA* **94**, 4516–4519.

19. Chakravarti, A. (1999) Population genetics: making sense out ofsequence. *Nat. Genet.* **21**, 56–60.

20. Collins, F. S., Guyer, M. S., and Chakravarti, A. (1997) Variations on a theme: cataloging human DNA sequence variation. *Science* **278**, 1580–1581.

21. Lander, E. S. (1996) The new genomics: global views of biology. *Science* **274**, 536–539.

22. Jorde L. B. (2000) Linkage disequilibrium and the search for complex disease genes. *Genome Res.* **10**, 1435–1444.

23. Moffatt, M. F., Traherne, J. A., Abecasis, G. R., and Cookson, W. O. (2000) Single nucleotide polymorphism and linkage disequilibrium within the TCR alpha/delta locus. *Hum. Mol. Genet.* **9**, 1011–1019.

24. Risch, N. and Merikangas, K. (1996) The future of genetic studies of complex human diseases. *Science* **273**, 1516–1517.

25. Kwok, P.-Y. (2000) Approaches to allele frequency determination. *Pharmacogenomics* **1**, 231–235.

26. Taillon-Miller, P., Bauer-Sardiña, I., Saccone, N. L., Putzel, J., Laitinen, T., Cao, A., et al. (2000) Juxtaposed regions of extensive and minimal linkage disequilibrium in human Xq25 and Xq28. *Nat. Genet.* **25**, 324–328.

27. Daly, M. J., Rioux, J. D., Schaffner, S. F., Hudson, T. J., and Lander, E. S. (2001) High-resolution haplotype structure in the human genome. *Nat. Genet.* **29**, 229–232.

28. Goldstein, D. B. and Weale, M. E. (2001) Population genomics: linkage disequilibrium holds the key. *Curr. Biol.* **11**, R576–R579.

29. Horikawa, Y., Oda, N., Cox, N. J., Li, X., Orho-Melander, M., Hara, M., et al. (2000) Genetic variation in the gene encoding calpain-10 is associated with type 2 diabetes mellitus. *Nat. Genet.* **26**, 163–175.

30. Reich, D. E., Cargill, M., Bolk, S., Ireland, J., Sabeti, P. C., Richter, D. J., et al. (2001) Linkage disequilibrium in the human genome. *Nature* **411**, 199–204.

31. Stam, L. F. and Laurie, C. C. (1996) Molecular dissection of a major gene effect on a quantitative trait: the level of alcohol dehydrogenase expression in *Drosophila melanogaster. Genetics* **144**, 1559–1564.

32. Templeton, A. R., Clark, A. G., Weiss, K. M., Nickerson, D. A., Boerwinkle, E., and Sing, C. F. (2000) Recombinational and mutational hotspots within the human lipoprotein lipase gene. *Am. J. Hum. Genet.* **66**, 69–83.

33. Wang, D. G., Fan, J.-B., Siao, C.-J., Berno, A., Young, P., Sapolsky, R., et al. (1998) Large-scale identification, mapping, and genotyping of single-nucleotide polymorphims in the human genome. *Science* **280**, 1077–1082.

34. Begun, D. J. and Aquadro, C. F. (1992) Levels of naturally occurring DNA polymorphism correlate with recombination rates in *D. melanogaster. Nature* **356**, 519–520.

35. Nachman, M. W. (1997) Patterns of DNA variability at X-linked loci in *Mus domesticus. Genetics* **147**, 1303–1316.

36. Nachman, M. W., Bauer, V. L., Crowell, S. L., and Aquadro, C. F. (1998) DNA variability and recombination rates at X-linked loci in humans. *Genetics* **150**, 1133–1141.

37. Charlesworth, D., Charlesworth, B., and Morgan, M. T. (1995) The pattern of neutral molecular variation under the background selection model. *Genetics* **141**, 1619–1632.

38. Rogers, A. R. and Harpending, H. (1992) Population growth makes waves in the distribution of pairwise genetic differences. *Mol. Biol. Evol.* **9**, 552–569.

39. McDonald, J. H., and Kreitman, M. (1991) Adaptive protein evolution at the *Adh* locus in *Drosophila. Nature* **351**, 652–654.

40. Hudson, R. R., Kreitman, M., and Aguade, M. (1987) A test of neutral molecular evolution based on nucleotide data. *Genetics* **116**, 153–159.

2

Denaturing High-Performance Liquid Chromatography

Andreas Premstaller and Peter J. Oefner

1. Introduction

Denaturing high performance liquid chromatography (dHPLC) is a fast and reliable technique for the DNA variation screening *(1,2)*. It can detect in minutes with close to 100% sensitivity and specificity single-base substitutions as well as small deletions and insertions in DNA fragments ranging from 80–1500 base pairs in size *(3,4)*. In partially denaturing HPLC, typically 2–10 chromosomes are compared as a mixture of PCR products. Upon mixing, denaturing and reannealing of amplicons containing one or more mismatches, not only the original homoduplices are formed again but, simultaneously, the sense and anti-sense strands of either homoduplex form two heteroduplices. Heteroduplices denature more extensively at elevated column temperatures in the range of 48–67°C; they are retained less on the chromatographic separation matrix, allowing the separation of homo- and heteroduplex species by ion-pair reversed-phase HPLC (IP-RP-HPLC) *(5)*. Characteristic peak patterns both for homozygous and heterozygous samples are obtained.

From: *Methods in Molecular Biology, vol. 212:*
Single Nucleotide Polymorphisms: Methods and Protocols
Edited by: P-Y. Kwok © Humana Press Inc., Totowa, NJ

In IP-RP-HPLC, the chromatographic phase system comprises a hydrophobic stationary phase and a hydroorganic eluent containing an amphiphilic ion and a small, hydrophilic counterion. An electrical potential is created at the surface of the hydrophobic stationary phase by adsorption of positively charged amphiphilic triethylammonium ions. Size dependent retention of DNA is governed by both the magnitude of individual surface potentials of stationary phase and sample and the contact area involved in electrostatic interaction. Upon increase of column temperature, the DNA double helix begins to denature partially, and forms a bubble that increases its outer surface and hence decreases the surface potential in the affected region of DNA. Consequently, retention is reduced. In the case of a mixture of homo- and heteroduplices, separation of all four species is primarily the result of differences in neighboring stacking interactions that determine the degree of destabilization.

DNA fragments shorter than approx 150 bp are too unstable to allow the detection of mutations by partially denaturing HPLC. However, the high-resolving power of IP-RP-HPLC enables the detection of mutations in short polymerase chain reaction (PCR) products (50–100 bp) under completely denaturing conditions *(6)*. The retention of single-stranded nucleic acids is sequence-dependent owing to the solvophobic interactions between the hydrophobic surface of the stationary phase and the hydrophobic nucleobases. Therefore, differences in base composition as small as a single base out of 100 bases suffice to separate two single stranded nucleic acids of identical size, and the alleles of a given polymorphic locus can be resolved without the addition of a reference chromosome. The only exception to this rule have been C to G transversions.

While dHPLC already offers high sensitivity and productivity, the miniaturization of the separation channel by using capillary columns of 50–320 µm inner diameter *(7)* is a prerequisite to achieve higher throughput, information content, and cost-effectiveness. The most significant advantage of capillary HPLC is the better signal height-to-sample mass ratio, as the peak concentration is proportional to the inverse square of the column diameter *(8)*. The same amount of an analyte will give in theory a signal that is approx 500 times higher with a 0.2 mm than a 4.6 mm column. In practice, this translates into significantly smaller injection volumes on the order of a few hundred nanoliters and significant savings can be accomplished in PCR reagent consumption.

The greater concentration sensitivity of the capillary format makes it possible to combine dHPLC with laser-induced fluorescence detection for SNP analysis in analogy to capillary electrophoresis in DNA sequencing, enabling higher throughput by color multiplexing *(9)*. In this technique, different amplicons are labeled with different fluorescence dyes during PCR using dye-labeled primers. The samples are analyzed simultaneously in one chromatographic column, and are monitored separately by observing their characteristic emission wavelengths.

Higher sample throughput can also be obtained by bundling of columns into arrays similar to those used in capillary electrophoresis, and using only one pump, injection, and detection device *(9)*. Finally, the volatile mobile phase components, low flow rate, and the on-line removal of cations from nucleic acid samples make IP-RP-HPLC highly suited for the direct coupling to electrospray ionization mass spectrometry (ESI-MS) *(10)*. Mass spectrometry will allow positive confirmation of the identity of the resolved components and the unambiguous genotyping of the amplified PCR fragments. Moreover, the identification even of heterozygous alleles eluting as one single chromatographic peak should become feasible because ESI-MS can directly analyze and deconvolute simple mixtures of nucleic acids in real matrices such as PCR reactions in a time frame of a few minutes *(11)*.

2. Materials

2.1. Polymerase Chain Reaction

The PCR mixture contains 10 mM Tris-HCl, pH 8.3, 50 mM KCl, 2.5 mM MgCl$_2$, 0.1 mM each of the four dNTPs, 0.2 µM of each primer, and 1 U of AmpliTaq Gold (Applied Biosystems, Foster City, CA) in deionized distilled water. For HPLC with laser-induced fluorescence detection, one of the primers is labeled with a fluorophore.

2.2. Sequences and Samples

For the examples shown in **Fig. 1** and **Fig. 2**, the following sequences, with priming regions typed in lower case and positions and chemical nature of polymorphic sites indicated in brackets, are used:

Sequence 1, 413 bp, gggggtataagtataaacaaaacTGACCCCATCG CTGCCCT CTTGGAGCTGAGAGTCTCATAAACAGCTTT A A G G T A A T A A A A T C A T T T T (C / A) T G T G CCACAGGATGTGAGTTGGTTTGATGACCCTAAAAACACC ACTGGAGCATTGACTACCAGGCTCGCCAATGATGCTGCT CAAGTTAAAGGGGTACGTGCCTCCTTTCTACTGGT(G/A) T T T G T C T T A A T T G G C (C / T) A T T T T G G A C C C C A G C A T G A A A C T A A T T T T C T C (C / A) T T A C G G G T G T T A G T T A T C A T C A T T A A G A A A A T G T T G A A T A A A T A T C T A A C C T A C G A A T A T A T C A C A T G C TTTTTGTAGCAACATGTTAACTATTTAAACATTATATACT GTAGAGCATATAGATAACTTATAAAccatttgctattgctgttatt; Sequence 2, 62 bp, cccaaacccattttgatgctT(G/T)ACTTAAaagg tctt caattattattt tcttaaatattttg. For the analysis with fluorescence detection (example in **Fig. 3**), sequences from the human *MDR1* gene containing one or two single nucleotide polymorphism are studied. The first is from exon 13, 380 bp in size, and amplified with HEX-labeled 5'ATCTTTCTGATGTTGCCCTTTC as forward and unla-beled 5'CCTTCTTAGGATTTCCCTTCTT as reverse primer. The second amplicon is 351 bp in length from exon 22, FAM-labeled 5'ACCACTATTTACTCTTGTGCCT and unlabeled 5'GTTCTA CCTTAGAGATGTCCCT are used as primers. Finally, a 327 bp fragment from exon 26 is amplified with 5'TGCTGAGAACAT TGCCTATGGAG as forward and 5'AACACTTTCATCCCTTCCT

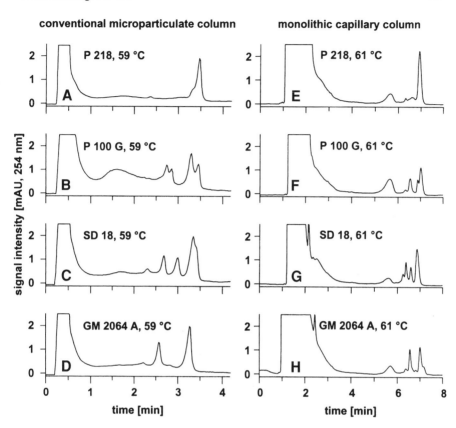

Fig. 1. Comparison of the chromatographic separation profiles obtained for one homozygous control and three heterozygotes under partially denaturing conditions using (**A–D**) a conventional microparticulate 4.6 mm ID column and (**E–H**) a monolithic 200 μm ID PS/DVB capillary column. Column, (**A–D**) 50 × 4.6 mm ID stainless-steel column packed with 2 μm PS/DVB-C18 particles (DNASep™, Transgenomic), (**E–H**) monolithic PS/DVB, 60 × 0.2 mm ID; mobile phase, (**A**) 100 m*M* TEAA, 0.1 m*M* Na₄EDTA, pH 7.0, (**B**) 100 m*M* TEAA, 0.1 m*M* Na₄EDTA, pH 7.0, 25% acetonitrile; linear gradient, (**A–D**) 50–52% B in 0.5 min, 52–59% B in 3.5 min, (**E–H**) 43–50% B in 0.5 min, 50–54% B in 3.5 min; flow rate, (**A–D**) 0.9 mL/min, (**E–H**) 3 μL/min; column temperature, (**A–D**) 59°C, (**E–H**) 61°C; injection volume, (**A–D**) 9 μL, (**E–H**) 500 nL; sample, sequence 1. *See* **Table 1** for the nature and location of mismatches underlying the chromatographic profiles. Adapted from **ref.** *(7)*.

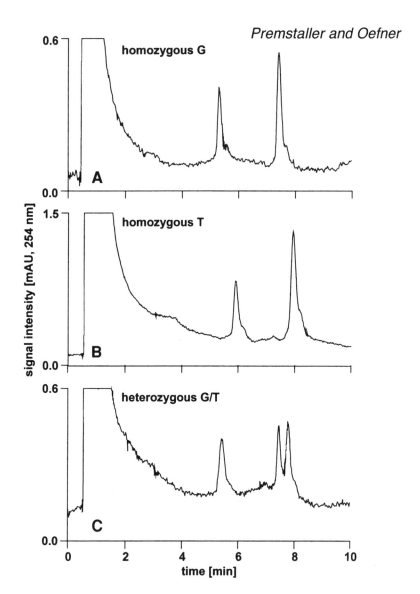

Fig. 2. Direct allelic discrimination of two alleles based on their different retention under completely denaturing conditions using a monolithic capillary column. Column, monolithic PS/DVB, 50 × 0.2 mm ID; (**A**) mobile phase, 100 m*M* TEAA, pH 7.0; (**B**) 100 m*M* TEAA, pH 7.0, 20% acetonitrile; linear gradient, 36–48% (**B**) in 10.0 min; flow rate, 3.0 µL/min; temperature, 75°C; detection, UV, 254 nm; injection volume, 500 nL; sample, sequence 2, (**A**) homozygous G, (**B**) homozygous T, (**C**) heterozygous G/T. Reproduced with permission from **ref.** *(7)*.

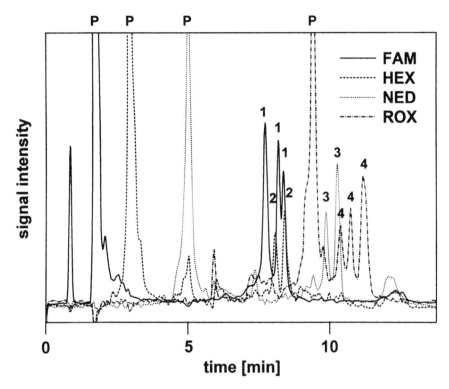

Fig. 3. Simultaneous analysis of 4 tagged amplicons from the MDR1-gene with capillary dHPLC and laser induced fluorescence detection. Column, PS/DVB monolith, 60 × 0.20 mm ID; mobile phase, 100 mM TEAA, 0.02 mmol/L Na$_4$EDTA, pH 7.0; linear gradient, 12.25–15.25% acetonitrile in 9 min; flow rate, 3.0 μL/min; temperature, 58°C; detection, excitation at 488 nm, emission measured at 525, 555, 580, and 590 nm; injection volume, 1 μL pooled sample, *MDR1*, (1) exon 22, 351 bp, FAM, (2) exon 13, 380 bp, HEX, (3) exon 26, 327 bp, ROX, (4) exon 26, 327 bp, NED.

CACA as reverse primer. To obtain NED-labeled amplicon, a NED-labeled forward primer is used together with an unlabeled reverse primer. For the ROX- labeled species, the reverse primer is labeled and forward primer is unlabeled. The mismatches on the four sequences are located at nucleotide position 291 (T/C) for exon 13, 320 (G/A) for exon 22, at position 212 (C/T) and 346 (T/C) for the

Table 1
**Position and Nature of Mismatches Contained
in 413 bp Amplicons of Three Heterozygous Individuals
and a Homozygous Control[a]**

Individual[b]	Positions of mutations from 5'-end of forward primer[c] (Positions within the *MDR1* sequence)			
	91 (2398-11)	209 (2481+24)	225 (2481+40)	258 (2481+73)
P218	C	G	C	C
P100G	C	**G/A**	C	C
SD18	**C/A**	G	C	C
GM2064A	C	**G/A**	**C/T**	**C/A**

[a]The sequence included exon 20 and flanking non-coding regions of the human P-glycoprotein (*MDR1*, Genbank Acc. No. M14758).
[b]Individuals are from the Stanford human diversity panel.
[c]Sequence 1, *see* Sequences and samples.

NED-labeled fragment and at position 325 (T/G) and 346 (T/C) for the ROX-labeled fragment from exon 26. Oligonucleotides labeled with the fluorescence dyes HEX, NED, and FAM are from Applied Biosystems, primers labeled with ROX and unlabeled primers are from Life Technologies (Rockville, MD). The *Hae*III digest of pUC18 is from Sigma (St. Louis, MO).

2.3. Instruments

1. A Perkin-Elmer 9600 thermal cycler (Applied Biosystems) is used in our laboratory. Other thermal cyclers with a heated lid are also suitable for this purpose. No mineral oil or wax overlay should be used.
2. The instrumentation for conventional HPLC consists of an on-line degasser (DG1210, Uniflows Co., Tokyo, Japan), two SD-200 high-pressure pumps, an electronic pressure module, a 600 µL dynamic mixer, a 6-port injection valve mounted into a MISTRAL column oven, an automated sample injector (AI-1A), a DYNAMAX UV-absorbance detector set at 254 nm, and a PC-based system controller and data analysis package (Varian Analytical Instruments, Walnut Creek, CA).

3. The capillary HPLC system comprises a low-pressure gradient micro pump (LCPackings, Amsterdam, Netherlands) controlled by a personal computer, a vacuum degasser (Uniflows Co.), a MISTRAL column oven, a six-port valve injector with a 1 μL sample loop (Valco Instruments Co., Houston, TX), a variable wavelength detector (UltiMate UV detector, LC Packings) with a Z-shaped capillary detector cell (UZ-LI-NAN, 3 nl cell, LC Packings), and a PC-based data system (UltiChrom, LC Packings).

4. Fluorescence data is collected using a four-color confocal fluorescence capillary array scanner *(12)*. Up to 25 capillaries in parallel can be mounted on a translation stage driven by a computer controlled microstepping indexer (Zeta6104, Compumotor Division of Parker Hannifin Co., Rohnert, CA). An excitation beam (488 nm) from an argon ion laser is focused into the capillary array through a microscope objective. Fluorescence from the capillaries is collected by the same objective and fractioned successively by four longpass dichroic beam splitters (Omega Optical, Brattleboro, VT) with transmission wavelengths (>50% T) of 505 nm, 540 nm, 570 nm, and 595 nm, respectively. Bandpass filters (Omega, 525DF30, 555DF30, 585DF20) are used on the light reflected by each of the first three beam splitters, and a 595 nm longpass filter after the last beam splitter. Filtered light in each fraction is then detected by a photomultiplier tube (Hamamatsu Corp., Bridgewater, NJ). Signals are lowpass filtered by 4-pole Bessel filters (824L8L-4, Frequency Devices, Haverhill, MA) and digitized by a 16-bit data acquisition board (CIO-DAS1402/16, Computer Boards Inc., Middleboro, MA). Data of four colors are filtered using a five-point filter and reduced using color-separation matrices. A computer program synchronizes the movement of the translation stage and data acquisition, allowing each capillary to be scanned at the rate of 2 Hz.

2.4. Chromatographic Columns for IP-IR-HPLC of Nucleic Acids

1. For conventional HPLC, the stationary phase consists of 2 μm micropellicular, alkylated PS/DVB particles *(13)* packed into 50 × 4.6 mm ID columns, which are commercially available (DNASep™, Transgenomic, San Jose, CA). Alternatively, Helix DNA columns (Varian) or Eclipse dsDNA Analysis Columns (Agilent Technologies, Waldbronn, Germany) are commercially available.

2. Monolithic capillary columns (60 × 0.2 mm ID) comprising a poly(styrene-*co*-divinylbenzene) stationary phase are prepared according to a previously published protocol *(14)*.

2.5. Eluents for IP-RP-HPLC of Nucleic Acids

A 2-*M* stock solution of triethylammonium acetate (TEAA), pH 7.0, can be obtained from Applied Biosystems (Foster City, CA) or prepared by dissolving equimolar amounts of triethylamine (Fluka, Buchs, Switzerland) and glacial acetic acid (Fluka) in water. High purity water is used for preparing the eluents.

1. Eluent A: 100 m*M* triethylammonium acetate, pH 7.0, 0.1 m*M* Na_4EDTA (Sigma).
2. Eluent B: 100 m*M* triethylammonium acetate, pH 7.0, 25% acetonitrile (HPLC grade, J. T. Baker, Phillipsburg, NJ), 0.1 m*M* Na_4EDTA.

The eluents can be stored up to 1 wk at room temperature.

3. Methods

3.1. Polymerase Chain Reaction

1. Polymerase chain reactions are performed in a 50 µL volume. 50 ng of genomic DNA are used for each single reaction.
2. The PCR cycling regime comprises an initial denaturation step at 95°C for 10 min to activate AmpliTaq Gold, 14 cycles of denaturation at 94°C for 20 s, primer annealing for 1 min at 63–56°C with 0.5°C decrements, and extension at 72°C for 1 min, followed by 20 cycles at 94°C for 20s, 56°C for 1 min, and 72°C for 1 min.
3. Following a final extension step at 72°C for 5 min, the samples are chilled to 6°C.

3.2. Formation of DNA Homo- and Heteroduplices

For dHPLC analysis, unpurified PCR products of each allele are mixed at an equimolar ratio and subjected to a 3 min 95°C denaturing step followed by gradual reannealing from 95°C to 65°C over 30 min. This ensures the formation of equimolar ratios of homo-

and heteroduplex species. The amplicons can be stored at 4°C for several weeks.

3.3. dHPLC Using Conventional Separation Columns

3.3.1. Conditioning and Testing of the Columns

1. New columns are conditioned in the conventional HPLC system with 50% eluent A and 50% eluent B using a flow rate of 0.9 mL/min at 50°C for 60 min.
2. Column performance is tested at a flow rate of 0.9 mL/min and 50°C by injecting 0.5 μg (300 fmol) of pUC18 *Hae* III restriction digest.
3. Elution is effected using linear gradients:

   0.0–3.0 min   43–56% eluent B
   3.0–10.0 min   56–68% eluent B

4. Afterwards, the column is washed with 95% B for 1 min.
5. Separation of the restriction fragments is monitored at the column outlet by UV absorbance at 254 nm.
6. Baseline resolution of the 257 bp and 267 bp, and the 434 bp and 458 bp fragments, respectively, should be obtained.

3.3.2. Chromatographic Analysis

For the reliable detection of mismatches, temperature and gradient conditions have to be chosen carefully, and special attention has to be paid to the thermal conditioning of the HPLC system (*see* **Notes 1** and **2**). The optimum temperature for the detection of mismatches can be determined either empirically (*see* **Note 3**) or, if the sequence of interest is known, by means of computation (*see* **Note 4**). Gradient start and end-points are proposed by the dHPLC-melt program (*see* **Note 4**), or are adjusted according to the size of the PCR products *(15)* based on the retention times of the DNA restriction fragments of the digest used to test column performance (*see* **Note 5**).

1. After equilibrating the HPLC system (*see* **Note 6**) at a flow rate of 0.9 mL/min using the initial gradient conditions with the column thermostat set to the optimum temperature for analysis, 10 μL of sample are injected.

2. Elution is effected using an acetonitrile gradient. To obtain the chromatograms shown in **Fig. 1A–D**, the following gradient program was applied:

   0.0–0.5 min 50–52% eluent B
   0.5–4.0 min 52–59% eluent B

3. Detection occurs with a UV absorbance detector at 254 nm.

4. After completion of the HPLC run, the column is washed with 95% eluent B for 1 min.

5. To maintain the performance of the chromatographic columns over several thousand injections, precautions are taken to avoid contamination of the chromatographic system with metal cations (*see* **Note 7**).

6. While not in use, the column is perfused at very low flow rate (0.05 mL/min) with 50% eluent A and 50% eluent B.

3.3.3. Evaluation of the Chromatograms

The formation of heteroduplices, and thus presence of mutations in the examined region of the chromosome are indicated by the appearance of more than one peak in the chromatogram. **Fig. 1A** shows the chromatographic analysis of a homozygous sample, where only one peak is observed. Different mutations (*see* **Table 1**) cause different degrees of destabilization at a given temperature and give rise to characteristic chromatographic profiles, as is depicted in **Fig. 1B–D**. However, the number of different profiles attainable is limited. Mutations located at the same nucleotide position always yield a different distinct chromatographic profile, while mutations located within the same melting domain tend to show very similar elution profiles *(16)*. Sequence analysis is still required to determine the exact location and nature of mismatches. However, it can be limited effectively to a few representative profiles. There appears to be no bias for specific mismatches, i.e., all possible single-base mismatches as well as insertions and deletions one to several base pairs in length are detected.

3.4. dHPLC Using Capillary Columns

3.4.1. Instrumental Requirements

When the chromatographic separation is downscaled from a conventional column of 4.6 mm ID to a miniaturized inner diameter of 0.2 mm, sample injection (*see* **Note 8**), gradient delivery (*see* **Note 9**), connective tubing (*see* **Note 10**), and detector (*see* **Note 11**) must be modified and optimized. Preheating of the mobile phase in the capillary HPLC system is obtained with a 20 cm 25 µm ID fused silica capillary positioned in the oven between the injector and the column.

3.4.2. Conditioning and Testing of the Columns

1. New monolithic columns are conditioned in the capillary HPLC system with 50% eluent A and 50% eluent B at a flow rate of 3 µL/min and 50°C for 60 min.
2. Column performance is tested at the same flow rate and temperature by injecting 3 ng (2 fmol) of pUC18 *Hae* III restriction digest using this gradient program:

   0.0–3.0 min   35–50% eluent B
   3.0–10.0 min 50–62% eluent B

3. Elution is followed by a washing step with 95% B for 0.5 min. Again, baseline resolution of the 257 bp and 267 bp, and the 434 bp and 458 bp fragments, respectively, should be obtained in the UV chromatogram.

3.4.3. Chromatographic Analysis

Using monolithic capillary column, slightly lower concentrations of acetonitrile are required for the elution of the amplicons compared to the conventional column. This is due to the more polar character of the PS/DVB surface of the monolithic capillary column compared to the conventional PS/DVB-C18 column. Since the double-helical structure of DNA is more stable at lower concentrations of organic modifier, a higher temperature has to be chosen to obtain partial denaturation. Usually an increase by 2°C is appropriate.

1. The HPLC system is equilibrated at a flow rate of 3 μL/min using the initial gradient conditions and analysis temperature.
2. To start the analysis, the gradient program is started at the pump.
   0.0–0.5 min   43–50% eluent B
   0.5–4.0 min   50–54% eluent B
3. However, injection of 500–1000 nL of the sample from PCR onto the capillary column occurs only after the gradient delay time has elapsed (*see* **Note 12**).
4. The recording of the chromatogram with a UV detector at 254 nm starts at the moment of injection.
5. After each run, the column is washed with 95% B for 1 min.

3.4.4. Evaluation of the Chromatograms

Figure 1E–H show chromatographic profiles acquired from the monolithic capillary column. The same four samples were used to obtain the profiles on the conventional microparticulate column depicted in **Fig. 1A–D**. The profiles are comparable and unambiguous identification of the heterozygous species is possible with both methods.

3.5. HPLC Using Completely Denaturing Conditions

3.5.1. Chromatographic Analysis

The same equipment described in the preceding sections is used for completely denaturing HPLC.

1. To ensure complete denaturation of the sample, the column temperature is set to 75°C.
2. After equilibration with 29% eluent B for 5 min at a flow rate of 3 μL/min, a linear gradient is started.
   0.0–10. min   29–39% eluent B
3. 500 nl of PCR sample are injected after the elapsing of the gradient delay time (*see* **Note 12**) and acquisition of the chromatogram is started with UV detection at 254 nm.

3.5.2. Evaluation of the Chromatograms

Figure 2 shows an example of allelic discrimination based on the separation of the single-stranded components of a 62 bp PCR product containing a single G to T transversion by means of dHPLC on a monolithic capillary column under completely denaturing conditions. The chromatograms in **Fig. 2A,B** illustrate two homozygous samples, where the two completely denatured single strands of the same chain length are completely separated by IP-RP-HPLC. In the case of a heterozygote sample, 3 peaks are observed corresponding to two coeluting and two separated single strands (*see* **Fig. 2C**). The single base change affects the retention of the single stranded components sufficiently to allow the discrimination of mutated from wild-type DNA in at least one pair of corresponding DNA strands.

Short primers with less than 20 nucleotides are often used to amplify the fragments used in completely denaturing HPLC. The decreased specificity of the primer annealing step can cause the amplification of additional sequences, that are detected as additional peaks in the chromatogram.

3.6. Multiplex Capillary dHPLC with Laser-Induced Fluorescence Detection

3.6.1. Sample Preparation

PCR of the homo- or heterozygous samples of interest is carried out separately with different fluorescently labeled primers for each sample. FAM, HEX, NED, and ROX have proven to be especially useful dyes for the chromatographic analysis due to their limited influence onto chromatographic retention. The high sensitivity of laser-induced fluorescence detection allows the pooling of up to four samples tagged with different dyes for dHPLC analysis (*see* **Note 13**).

3.6.2. Chromatographic Analysis

1. The capillary HPLC system is connected to the laser-induced fluorescence detector.
2. The HPLC system is equilibrated at a flow rate of 3 µL/min using the initial gradient conditions at 58°C.
3. After injection of 1 µL of pooled sample, the analytes are eluted using the following gradient:

    0.0–10.0 min   49–61% eluent B

4. The emission is monitored at four wavelengths. After each run, the column is washed with 95% B for 1 min.

3.6.3. Evaluation of the Chromatograms

Figure 3 shows the multiplex analysis and detection of four different heterozygous PCR products from the *MDR1* gene labeled with four different fluorescent dyes in one single chromatographic run. Species tagged with different dyes are identified at the characteristic emission wavelength of each attached fluorophore measured in the four channels of the detector. The emission maxima are at 525 nm for FAM, 555 nm for HEX, 580 nm for NED, and 605 nm for ROX. Spectral overlap of the emission spectra is corrected by means of a color separation matrix deduced from the emission spectra of the dye-labeled primers. The four peaks indicated with P stem from labeled primers, the peaks of the amplicons are indicated with 1 (FAM), 2 (HEX), 3 (NED), and 4 (ROX). For all four colors, the primers are well separated from the amplicon peaks, and the characteristic chromatographic profiles allow the assignment of the samples as heterozygous.

3.6.4. Influence of Fluorescence Dyes on Chromatographic Retention and Profile

The fluorophores are large apolar molecules. Additional solvophobic interactions between the apolar stationary phase and the fluorophores lead to an increased retention. The affinity to the column determined by retention time is, in increasing order, FAM < HEX < NED < ROX. A higher percentage of organic solvent is required for elution of the dye-labeled samples. Higher acetonitrile concentrations lower the temperature required to denature a given sequence, thus the chromatographic elution profiles seen with increasing hydrophobicity of the fluorophore resemble that of unlabeled fragments at increasing column temperature. At 58°C, FAM- and HEX-labeled PCR products yield heteroduplex profiles similar to those of unlabeled amplicons acquired at the same temperature. Profiles obtained from amplicons labeled with NED and ROX resemble those of corresponding unlabeled fragments analyzed at 1–2°C higher temperatures, respectively. Only fluorophores with similar effects on retention should be chosen in order to ensure maximum mismatch detection sensitivity. Alternatively, fragments with higher melting characteristics can be selected to compensate for the drop in melting temperature as a function of the fluorophore attached.

4. Notes

1. To obtain baseline resolution for heterozygous species, the DNA has to be preheated online for 2.5 s in a heat exchanger made of 80 cm of 250 μm ID PEEK tubing encased into a tin alloy block (HEX-440.010, Timberline Instruments, Boulder, CO) that is placed in front of the sample loop in the oven.
2. Direct contact between the stainless-steel column and the hot metal surfaces of the column oven must be avoided. The surfaces are warmer than the circulating air in the oven. Hence, direct contact will heat the mobile phase in the column to a higher temperature than indicated on the column oven display. Such contact will result in reduced reproducibility and discrepancies between predicted and observed temperatures.
3. The optimum temperature may range from 48–68°C for very AT- and GC-rich sequences, respectively. To empirically determine the optimum temperature at which to screen a particular sequence, a test sample is injected repeatedly at gradually increasing column temperatures until the duplex product peak begins to shift significantly (~1 min) towards shorter retention times. At this point, the presence of a mismatch will be usually detected by the appearance of one or two additional peaks eluting immediately before the homoduplex signal. Special care has to be taken that no low-melting domains are missed. Mutations may go undetected due to complete denaturation of such domains.
4. A melting algorithm has been developed and is freely available at the website http://insertion.stanford.edu/melt.html. Its use is recommended whenever sequence information is available. The site melting temperatures are defined as the temperatures at which the sites

are closed in 50% of fragments. Analysis should be performed at the highest temperature recommended. In case that the program calculates melting temperatures spanning more than 3–4°C, it is recommended to repeat the analysis at a temperature 3–4°C lower than the highest temperature recommended. The program also proposes chromatographic gradient conditions for the analysis of the sample.

5. The gradient is started 7% below the estimated percentage of eluent B at which the amplicon is expected to elute at 50°C, in order to account for the shift in retention towards lower percentages of eluent B with increasing column temperature. Once the optimum temperature has been determined, a gradient window as small as 4.5% over 2.5 min can be set. For amplicons up to 600 bp, the gradient is increased by 1.8% buffer B per minute. For larger amplicons, a more shallow gradient, e.g., 1.2% B per minute is used.

6. Equilibration time is dependent on the dead volume of the chromatograph, i.e., the volume of the liquid pathway between solvent mixer and column outlet. For equilibration between runs, the dead volume is at least replaced once, which typically takes 1–3 min.

7. Formation of rust in those parts of the liquid pathway that get in contact with the DNA, especially the frits at the column in- and outlets, and contamination of the separation column with metal cations dramatically decreases column performance. Therefore, all liquid contact parts should be made preferably of PEEK or titanium. The addition of 0.1 mM EDTA to both eluents appears to be equally effective. A column contaminated with metal cations can be rescued by repeated injections of 30–50 μL of 0.5 M Na$_4$EDTA.

8. To introduce sample onto a capillary column of 50–320 μm ID, the HPLC instrument is configured with a low dispersion valve for the injection of typically 20 nL to 2 μL of sample.

9. Reproducible gradients at low flow rates from 100 nL/min to 5 μL/ min are conveniently obtained by splitting a relatively high primary flow of mobile phase delivered by a gradient pumping system by means of a T-piece *(17)*. Thus, only a small portion of the mobile phase is passed via the injector onto the separation column, whereas the main flow of the mobile phase goes to waste. Since the primary flow is usually in the range of 100–250 μL/min, a reduction by a factor of 4–10 in solvent consumption is usually feasible compared to conventional 4.6 mm ID columns operated at a flow rate of 1 mL/min.

10. An excess of extra-column volume causes peak asymmetry and broadening. To retain the chromatographic resolution and minimize the delay volume, the volume of connecting tubing in the HPLC system must be minimized. Fused silica capillary tubing of an inner diameter not exceeding 25 μm has proven to be highly suited as a connection material, and all lines should be trimmed to as short a length as possible. A further reduction of the extra-column volume to less than 10% of the peak retention volume can be achieved by positioning of injection valve and sample loop in the column oven.

11. For optical detection, a sensitive detector with a low-volume detection cell of 1–60 nL is required *(18)*.

12. The low flow rate in microscale gradient HPLC separations is the cause of the so-called gradient delay time, the time passing between formation of the gradient in the mixing system of the pump and the arrival of the gradient on the separation column. The lower the flow rate, the more time is needed to flush the volume between the mixing device and the separation column. The gradient delay time of the HPLC system is determined experimentally following these steps:

 a. Set the flow rate to the desired value.

 b. Equilibrate the system at the expected operating temperature.

 c. Using UV detection at 215 nm, determine the elution time of an unretained sample t_0 by injecting a mixture of 0.1% acetone in water onto the column.

 d. Record the detector baseline at 215 nm of the following pump program, and determine the time t_B of the steep ascent of the baseline:

 0.0–5.0 min   5% eluent B
 5.0–5.1 min   5–95% eluent B
 5.1–10.0 min   95% eluent B

 e. Calculate the gradient delay time t_{Gr}: $t_{Gr} = t_B - 5$ min $- t_0$
 The gradient delay time varies with the temperature at which the chromatographic column is operated. It should be newly determined after modifications of components of the chromatographic system, especially replacement of tubing or the chromatographic column, or if the operating temperature is changed by more than 10°C.

13. The sensitivity of fluorescence detection is by a factor of 30–100 higher than that of UV absorbance detection *(19)*. All tested fluorescent dyes had comparable emission levels, the lowest from ROX is about 50% of the highest from FAM.

References

1. Oefner, P. J. and Underhill, P. A. (1995) Comparative DNA sequencing by denaturing high-performance liquid chromatography (dHPLC). *Am. J. Hum. Genet.* **57 (Suppl.)**, A266.
2. Xiao, W. and Oefner, P. J. (2001) Denaturing high performance liquid chromatography: a review. *Human Mutat.* **17**, 439–474.
3. Jones, A. C., Austin, J., Hansen, N., Hoogendoorn, B., Oefner, P. J., Cheadle, J. P., and O'Donovan, M. C. (1999) Optimal temperature selection for mutation detection by denaturing HPLC and comparison to single-stranded conformation polymorphism and heteroduplex analysis. *Clin. Chem.* **45**, 1133–1140.
4. Spiegelman, J. I. , Mindrinos, M. N., and Oefner, P. J. (2000) High-accuracy DNA sequence variation screening by dHPLC. *Biotechniques* **29**, 1084–1092.
5. Oefner, P. J. & Underhill, P. A. (1998) DNA mutation detection using denaturing high-performance liquid chromatography, in *Current Protocols in Human Genetics* (Dracopoli, N. C., Haines, J. L., Korf, B. R., et al., eds.), John Wiley & Sons, New York, NY, pp. 7.10.1–7.10.12.
6. Oefner, P. J. (2000) Allelic discrimination by denaturing high-performance liquid chromatography. *J. Chromatogr.* **B 739**, 345–355.
7. Huber, C. G., Premstaller, A., Xiao, W., Oberacher, H., Bonn, G. K., and Oefner, P. J. (2001) Mutation detection by capillary denaturing high-performance liquid chromatography using monolithic columns. *J. Biochem. Biophys. Methods* **47**, 5–19.
8. Ishii, D. (1988) *Introduction to Microscale High-Performance Liquid Chromatography.* VCH, Weinheim.
9. Premstaller, A., Xiao, W., Oberacher, H., et al. (2001) Temperature-modulated arrary high-performance liquid chromatography. *Genome Res.* **11**, 1944–1951.
10. Huber, C. G. and Krajete, A. (1999) Analysis of nucleic acids by capillary ion-pair reversed-phase HPLC coupled to negative ion-electrospray ionization mass spectrometry. *Anal. Chem.* **71**, 3730–3739.
11. Oberacher, H., Oefner, P. J., Parson, W., Huber, C. G. (2001) On-line liquid chromatography-mass spectrometry: A useful tool for the detection of DNA sequence variation. *Angew. Chem. Int. Ed.* **40**, 3828–3830.

12. Kheterpal, I., Scherer, J. R., Clark, S. M., Radhakrishnan, A., Ju, J., Ginther, C. L., et al. (1996) DNA sequencing using a four-color confocal fluorescence capillary array scanner. *Electrophoresis* **17**, 1852–1859.
13. Huber, C. G., Oefner, P. J., and Bonn, G. K. (1993) High-resolution liquid chromatography of oligonucleotides on highly crosslinked poly(styrene-divinylbenzene) particles. *Anal. Biochem.* **212**, 351–358.
14. Premstaller, A., Oberacher, H., and Huber, C. G. (2000) High-performance liquid chromatography-electrospray ionization mass spectrometry of single- and double stranded nucleic acids using monolithic capillary columns. *Anal. Chem.* **72**, 4386–4393.
15. Huber, C. G., Oefner, P. J., and Bonn, G. K. (1995) Rapid and accurate sizing of DNA fragments by ion-pair reversed-phase chromatography on alkylated nonporous poly-styrene/divinylbenzene particles. *Anal. Chem.* **67**, 578–585.
16. O'Donovan, M. C., Oefner, P. J., Roberts, C. S., Austin, J., Hoogendoorn, B., Guy, C., et al. (1998) Blind analysis of denaturing high-performance liquid chromatography as a tool for mutation detection. *Genomics* **52**, 44–49.
17. Chervet, J. P. (1991) Micro flow processor. EP 0495255A1.
18. Chervet, J. P., Ursem, M., and Salzmann, J. P. (1996) Instrumental requirements for nanoscale liquid chromatography. *Anal. Chem.* **68**, 1507–1512.
19. Oefner, P. J., Huber, C. G., Umlauft, F., Berti, G. N., Stimpfl, E., and Bonn, G. K. (1994) High-resolution liquid chromatography of fluorescent dye labeled nucleic acids. *Anal. Biochem.* **223**, 39–46.

3

SNP Detection and Allele Frequency Determination by SSCP

Tomoko Tahira, Akari Suzuki, Yoji Kukita, and Kenshi Hayashi

1. Introduction

Single-strand conformation polymorphism (SSCP) analysis is a sensitive mutation detection system that has been widely used in the field of medical genetics *(1,2)*. In this method, PCR products are denatured to become single-stranded, and separated by gel electrophoresis under nondenaturing conditions. A single-stranded fragment with a mutation or single nucleotide polymorphism (SNP) has a different conformation from its wild-type counterpart, and these conformational differences result in differing electrophoretic mobility. To identify SNPs at polymorphic sequence-tagged sites (STSs), it is necessary to sequence the STSs in individuals with different genotypes. However, once an SNP sequence is correlated with the corresponding fragment mobility in an SSCP analysis, sequencing may not be necessary for genotyping, because SSCP electrophoresis is highly reproducible *(3,4)*.

For large-scale SNP analysis, we have developed a semi-automated and streamlined method, PLACE-SSCP, in which polymerase chain reaction (PCR) products are postlabeled with two

From: *Methods in Molecular Biology, vol. 212:*
Single Nucleotide Polymorphisms: Methods and Protocols
Edited by: P-Y. Kwok © Humana Press Inc., Totowa, NJ

different fluorescent dyes in one tube, and analyzed by automated capillary electrophoresis under SSCP conditions *(4,5)*. In this method, no synthesis of fluorescent PCR primers is required *(6)*. The use of a capillary-based automated sequencer allows precise control of electrophoretic conditions. Because the machine can detect four different fluorophores, mobility calibration using internal standards is also possible *(4,7)*.

The advantage of PLACE-SSCP in SNP analysis is that it can be used to estimate the allele frequencies of SNPs from an analysis of pooled DNA. SNP alleles are separated as peaks on the electropherogram, and their frequencies can be reliably and accurately quantified from the peak heights. A cost-effective method for the estimation of allele frequencies of SNPs, such as that described here, is required before these candidate markers can be used for large-scale genetic studies, because there are now millions of candidate SNPs in public databases, and choosing informative SNPs from such a collection is a considerable task *(8)*.

In this chapter, we describe a strategy with which to quantify allele frequencies of some candidate SNPs from a public database. SNPs of moderate to high heterozygosity (minor allele frequencies greater than 10%) can be efficiently detected, and their allele frequencies accurately estimated by PLACE-SSCP analysis of pooled DNA samples *(9)*.

2. Materials

In this chapter, the materials and methods required specifically for PLACE-SSCP are described. Protocols for direct sequencing of PCR products are available from the suppliers of sequencing kits, and are not included here.

2.1. DNA Samples

1. The concentration of each DNA sample is estimated by UV spectrophotometry in two independent assays. Each sample is diluted to a concentration of 50 µg/mL (*see* **Note 1**).

2. To make a sample of pooled DNA, equal amounts of DNA from each individual sample are manually combined.

2.2. PCR

1. Primers carrying the sequences 5'-ATT-3' or 5'-GTT-3' at their 5' ends are synthesized, as described previously *(4,5)*. They are diluted with 0.1X TE to a final concentration of 2.5 μ*M* and stored frozen.
2. An equivolume (1:1 v/v) mixture of Ampli*Taq* DNA polymerase (5 U/μL; Applied Biosystems, Foster City, CA) and *Taq*Start antibody (1.1 mg/mL; Clontech, Palo Alto, CA) is prepared as recommended by the supplier of the antibody, and stored at –20°C.
3. 25 m*M* MgCl$_2$.
4. 10X PCR buffer: 0.5 *M* KCl, 0.1 *M* Tris-HCl, pH 8.3.
5. 1.25 m*M* dNTP: a mixture of an 1.25 m*M* of each of dATP, dCTP, dGTP, and dTTP (Amersham Pharmacia, Piscataway, NJ), stored at –20°C.

2.3. Post-PCR Fluorescent Labeling

1. Four fluorescent nucleotide stock solutions, i.e., 100 μ*M* R6G-dCTP, 100 μ*M* R110-dUTP, 400 μ*M* TAMRA-dCTP, and 400 μ*M* TAMRA-dUTP (Applied Biosystems), stored at –20°C.
2. 10X Klenow buffer: 50 m*M* Tris-HCl, pH 8.7, 0.1 *M* MgCl$_2$.
3. 0.2 *M* Na$_2$EDTA, pH 7.8.
4. DNA polymerase I Klenow fragment: 5 U/μL (New England Biolabs, Beverly, MA), stored at –20°C.
5. Calf intestine alkaline phosphatase: 20 U/ml enzyme (Roche, Indianapolis, IN), stored at 4°C, diluted to 0.2 U/μL using 1X CIP buffer (50 m*M* Tris-HCl, pH 8.5, 0.1 m*M* EDTA) just before use.

2.4. Capillary Electrophoresis

1. Automated capillary sequencer: ABI PRISM 310 (Applied Biosystems) and software (ABI PRISM 310 Data Collection Software, version 1.0.4 or higher, ABI Prism Run Module, GS Template, GeneScan Analysis Software, version 2.0.2 or higher). An SSCP analysis matrix file should be prepared (*see* **ref. *10***: GeneScan Reference Guide, for directions on preparing a matrix file).

2. Uncoated fused silica capillaries, each with a total length of 41 cm (effective separation length = 36 cm) and an internal diameter of 50 μm (Applied Biosystems).
3. 1X TBEG buffer, pH 7.8: 90 mM Tris base, 90 mM boric acid, 2 mM Na$_2$EDTA, 10% glycerol.
4. Size marker: GeneScan 500 TAMRA (Applied Biosystems), purified by ultrafiltration using Microcon 50 (Millipore, Bedford, MA) to remove ions and low molecular-weight fragments, and adjusted to half its original volume.
5. Deionized formamide (stored at –20°C in aliquots).
6. Separation matrix: GeneScan Polymer (Applied Biosystems) in 1X TBEG. GeneScan Polymer is supplied as a 7% solution, which is diluted to 6% by mixing with 20X TBE and glycerol (*see* **Note 2**). It is important to prevent the generation of air bubbles in the polymer during the run, especially when a polymer solution of high viscosity is used. For this purpose, we treat the polymer solution under vacuum for 5 min and then centrifuge it for 20 min at 3000g immediately before use.

3. Methods

3.1. PCR

1. Amplify the target sequence by PCR using one primer with ATT at its 5' end, and the other with GTT at its 5' ends (final primer concentrations of 0.25 μM each, and nucleotide concentrations of 200 μM each). DNA from individuals and pooled DNA samples are used as templates (*see* **Note 3**). Typically, the amplification is by 40 cycles of PCR using the template at 5 ng/μL.
2. Confirm specific amplification by agarose electrophoresis. It is important to optimize the PCR so that few or no fortuitous fragments are produced in the amplification reaction, especially when the primers are newly synthesized.

3.2. Fluorescent Labeling Using a 3' Exchange Reaction

1. Combine the following solutions to make 50 μL of labeling mix (for twelve labeling reactions): 10 μL 10X Klenow buffer, 2 μL 100 mM R110-dUTP, 2 μL 100 μM R6G-dCTP, 2 μL Klenow fragment

(20 U), and 34 µL water. For the reference sample, use 2 µL 400 µ*M* TAMRA-dCTP and 2 µL 400 µ*M* TAMRA-dUTP instead of R110-dUTP and R6G-dCTP.

2. Add 4 µL labeling mix to each PCR tube containing 4 µL amplified product, vortex, and incubate for 30 min at 37°C.
3. Stop the action of the Klenow fragment by adding 0.8 µL 0.2 *M* EDTA, then add 10 µL calf intestine alkaline phosphatase (0.2 U/µL); vortex; and incubate for 30 min at 37°C to degrade the nucleotides (*see* **Note 4**).
4. Store at 4°C until use.

3.3. Electrophoresis

1. Create the run module for the SSCP (*see* **ref. *10***: ABI GeneScan Reference Guide), with the following settings: injection time, 5–10 s; injection voltage, 15 kV; collection time, 20–35 min; EP voltage, 15 kV; heatplate temperature, 30°C (*see* **Notes 5,6**).
2. Complete the sample sheet and injection list of the ABI PRISM 310 Data Collection Software.

3.4. Sample Loading

1. Mix postlabeled PCR products (sample and reference, 0.5 µL each) with 0.5 µL purified GS500-TAMRA and 13.5 µL formamide. Incubate the mixture at 95°C for 5 min.
2. Place the sample tray on the autosampler and start the run.

3.5. Identification of SNP Alleles

1. After electrophoresis under SSCP conditions, analyze the raw data using the GeneScan Analysis software. Normalize run-to-run variation in retention times using one arbitrarily chosen electropherogram as template. To do this, align the peaks of the internal control (GS500-TAMRA and reference, or GS500-TAMRA alone) in the remaining electropherograms with those of the templates. Sample peak positions between the peaks of the internal control are calibrated by the "Local Southern" interpolation method with reference to the peaks of the internal standards *(4,7)*.
2. Compare the peak patterns of different individuals and identify peaks that are absent in some individuals (*see* **Note 7**). Genotype individu-

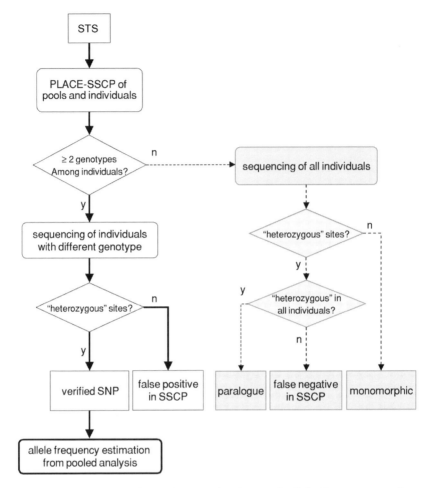

Fig. 1. Flow chart of the characterization and allele frequency estimation of candidate SNPs. Estimation of allele frequencies of SNPs detectable by PLACE-SSCP is performed as indicated by thick solid lines. The flow indicated by dotted lines is not required if the purpose of the experiment is to determine the allele frequencies of informative SNPs.

als as heterozygotes or homozygotes according to these allelic peaks. If all individuals show the same peak pattern, several possible cases should be considered (*see* **Note 8**, **Fig. 1**).

3. Select at least one heterozygote and one homozygote, and examine the polymorphisms by direct sequencing. Polymorphic nucleotides are identified by PolyPhred analysis *(11)* and visual inspection.

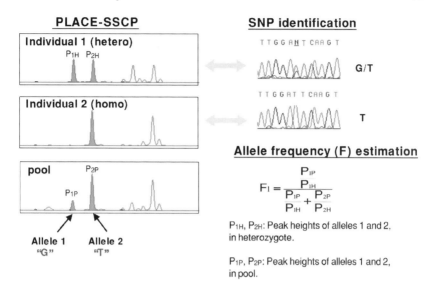

Fig. 2. Examples of PLACE-SSCP analysis of pooled and individual DNA samples. Electropherograms from PLACE-SSCP analyses of two individuals and a pooled sample are shown on the left. Sequencing traces for the individuals are shown on the right. Arrows indicate the positions of SNPs. Allele frequencies are calculated by the equation indicated.

4. Correlate which peak of the single-stranded DNA corresponds to which SNP allele (SNP sequence).

3.6. Quantification of Allele Frequency

1. Record the peak height of each allele onto a spreadsheet, e.g., Microsoft Excel.
2. Calculate the allele frequency using the equation:

$$F_i = \frac{P_i/H_i}{\sum_{j=1}^{n} P_j/H_j} \tag{1}$$

where n is the number of alleles of the STS, F_i is the frequency of the ith STS allele, P_i is the peak height of the ith allele in the pool, and H_i is the correction factor. H_i is the relative peak height of the alleles when they are present in equimolar ratios, and is calculated from the peak height ratios in heterozygotes of all combinations (*see* **Note 9**, **Fig. 2**).

4. Notes

1. It is important that pooled DNA is made of equal parts of individual DNAs. When the amount of DNA is limited, it should be quantified by a fluorescence-based assay using the intercalating dye PicoGreen (Molecular Probes Inc., Eugene, OR), and measured using a fluorescence plate reader. This method is also useful in quantifying a large number of samples. See the manufacturers' instruction for details.

2. Nondenaturing polymers should be used for SSCP. We used Performance Optimized Polymer (POP) without urea (kindly provided by Applied Biosystems) in the previous study *(9)*. Alternatively, dimethylacrylamide polymer can be used for SSCP *(12)*.

3. Alternatively, several (e.g., 8 –12) individuals are first genotyped by SSCP. One heterozygote and one homozygote are then selected and analyzed using SSCP together with the pooled DNA samples to determine allele frequencies.

4. Unincorporated fluorescent dNTPs can also be removed by gel filtration. We use Multiscreen HV plates (Millipore) with Sephadex G-50 superfine (Amersham Pharmacia), when multiple samples are processed. Gel filtration has the advantage of removing the salts in the reaction mixture. Samples purified by this method can be injected after dilution with 0.5 mM EDTA without the addition of formamide.

5. Electrophoresis under SSCP conditions is sensitive to temperature. Therefore, temperature control is essential for reproducible results. ABI PRISM 310 controls temperatures between ambient and 60°C. To maintain the run temperature at 30°C, it is important to keep the ambient temperature at 25°C with air conditioning.

6. The migration time of the peak is recorded as data points, and is usually set to 220 ms/data point. The mode of data collection can be modified to 50 ms of integration time followed by 20 ms of idle time (70 ms/datapoint) to improve resolution. Consult agents of Applied Biosystems for this modification.

7. In almost all cases, the peaks of two complementary strands (labeled differentially with R6G or R110) migrate differently. In many cases, only one of the two strands shows a clear mobility shift. The peak of the strand that shows better separation of the alleles is used for SNP typing and the determination of allele frequencies. Peaks of fluorescence showing both colors that migrate faster than single-stranded peaks are presumably the peaks of double strands or aggregated strands. They do not interfere with the analysis.

8. The strategy we use to examine the efficiency of PLACE-SSCP in SNP verification is shown in **Fig. 1**. When the same electropherogram is obtained for all individuals, we sequence the STS of all individuals. In our experience of analyzing 194 STSs (80–400 bp each), less than 10% of SNPs went undetected by SSCP (false negatives). No false-positives were found. When heterozygous nucleotides were found among all the individuals examined, amplification of paralogous sequence was inferred.

9. In many cases, the two alleles of the heterozygote do not give peaks of equal height. This is most likely attributable to biased PCR amplification of the alleles. In the given equation, a difference in peak height between alleles is normalized using heterozygous individuals. The peak height ratio of the heterozygote is also used as a measure of the reproducibility of quantification by SSCP, and should be checked before the determination of actual allele frequencies using pooled DNA.

PLACE-SSCP method using capillary array sequencer, such as ABI3100, ABI3700 (Applied Biosystems), and MegaBace (Amersham Pharmacia) has been developed in our laboratory and will be published elsewhere.

References

1. Orita, M., Suzuki, Y., Sekiya, T., and Hayashi, K. (1989) Rapid and sensitive detection of point mutations and DNA polymorphisms using the polymerase chain reaction. *Genomics* **5**, 874–879.
2. Hayashi, K. and Yandell, D. W. (1993) How sensitive is PCR-SSCP? *Human Mutation* **2**, 338–346.
3. Hayashi, K. (1999) Recent enhancements in SSCP. *Gen. Anal. Biomol. Eng.* **14**, 193–196.
4. Inazuka, M., Wenz, H. M., Sakabe, M., Tahira, T., and Hayashi, K. (1997) A streamlined mutation detection system: multicolor post-PCR fluorescence-labeling and SSCP analysis by capillary electrophoresis. *Genome Res.* **7**, 1094–1103.
5. Hayashi, K., Kukita, Y., Inazuka, M., and Tahira, T. (1998) Single strand conformation polymorphism analysis in: *Mutation Detection: A Practical Approach*, Cotton, R. G. H., Edkins, E., and Forrest, S., eds., Oxford University Press, Oxford, UK, pp. 7–24.

6. Inazuka, M., Tahira, T., and Hayashi, K. (1996) One-tube post-PCR fluorescent labeling of DNA fragments. *Genome Res.* **6**, 551–557.

7. Hayashi, K., Wenz, H.-M., Inazuka, M., Tahira, T., Sasaki, T., and Atha, D. H. (2001) SSCP analysis of point mutations by multicolor capillary electrophoresis, in *Capillary Electrophoresis of Nucleic Acids*, vol. 2 (Michelson, K. R. and Cheng, J., eds.), Humana Press, Totowa, NJ, pp. 109–126.

8. Marth, G., Yeh, R., Minton, M., Donaldson, R., Li, Q., Duan, S., et al. (2001) Single-nucleotide polymorphisms in the public domain: how useful are they? *Nat. Genet.* **27**, 371–372.

9. Sasaki, T., Tahira, T., Suzuki, A., Higasa, K., Kukita, Y., Baba, S., and Hayashi, K. (2001) Precise estimation of allele frequencies of single-nucleotide polymorphisms by a quantitative SSCP analysis of pooled DNA. *Am. J. Hum. Genet.* **68**, 214–218.

10. GeneScan Reference Guide: Chemistry Reference for the ABI Prism 310 Genetic Analyzer, http://docs.appliedbiosystems.com/pebiodocs/04303189.pdf

11. Nickerson, D. A., Tobe, V. O., and Taylor, S. L. (1997) PolyPhred: automating the detection and genotyping of single nucleotide substitutions using fluorescence-based resequencing. *Nucleic Acids Res.* **25**, 2745–2751.

12. Ren, J., Ulvik, A., Refsum, H., and Ueland, P. M. (1999) Application of short-chain polyacrylamide as sieving medium for the electrophoretic separation of DNA fragments and mutation analysis in uncoated capillaries. *Anal. Biochem.* **276**, 188–194.

4

Conformation-Sensitive Gel Electrophoresis

Arupa Ganguly

1. Introduction

Several large databases are now available which contain information on hundreds of thousands of single nucleotide polymorphisms (SNPs) distributed throughout the genome. Although these databases represent a tremendous resource for studies of human variation and disease, two challenges remain. First is the development of novel strategies of genotyping known SNP based genetic markers that will allow accurate and rapid high-throughput analyses *(1,2)*. Second is the development of sensitive and specific methods for the detection of previously unknown, novel SNPs in particular genes or genomic regions of special interest. The various methods currently available for detection of SNPs all depend on the ability to detect different physical properties in DNA molecules that result from variations in the nucleotide sequence. These properties include minor differences in thermal melting profiles of two DNA molecules differing in sequence by a single base or structural distortions in perfectly double stranded nucleic acid molecules due to the presence of unpaired or mismatched bases.

From: *Methods in Molecular Biology, vol. 212:*
Single Nucleotide Polymorphisms: Methods and Protocols
Edited by: P-Y. Kwok © Humana Press Inc., Totowa, NJ

Nuclear magnetic resonance (NMR) studies and X-ray analysis of DNA molecules containing unpaired or mismatched bases demonstrate that any structural alteration, if present, gives rise to very subtle measurable changes *(3)*. Under some conditions, presence of an extra unpaired base on one strand of a double stranded DNA molecule can produce a large change and bend of about 20° *(4)*. However, NMR and X-ray diffraction methods require highly sophisticated instrumentation and can not be easily accessible to every research laboratory for individual genotyping projects.

A different set of methods explore the possibility that single base differences in DNA sequences can be detected by differential migration of single stranded molecules containing variant DNA sequences (SSCP) or double stranded molecules consisting of heteroduplexes and homoduplexes in electrophoretic gels (CSGE). These methods require minimal manipulation of the PCR amplified genetic materials and are very useful in terms of easy access, cost and high throughput in scanning of large regions of genomic DNA or cDNA for presence of sequence variation *(5)*.

1.1. Single-Strand Conformation Polymorphism Analysis

In recent years, single-strand conformation polymorphism (SSCP) has been one of the most frequently used methods for identifying single base mutations in many putative disease causing genes *(6)*. In this method the PCR products are denatured followed by rapid cooling such that the complimentary DNA strands fold back on themselves and acquire specific secondary structures defined by the nucleotide sequence of the fragment to be analyzed. When these fragments are analyzed in a nondenaturing polyacrylamide gel, differential migration is observed for complimentary strands of the same DNA molecule containing sequence variation as small as a single base. This method has found its biggest application in rapid and preliminary survey of large sets of samples to determine a reasonable estimate of the frequency of a previously known mutation or polymorphism. The use of radioactivity enhances the sensitivity of this method but non-radioactive detection methods can be used as well that include fluorescently labeled DNA *(7)* and silver staining *(8)*. However SSCP method can have its own limitations when used for screening large number of SNPs. There is no set of simple rules for the choice of PCR primers based on the DNA sequence that can predict optimized separation for all SNPs. Usually one has to test at least four different combinations of gel electrophoresis conditions including gel temperature (room temperature or colder) and the presence or absence of glycerol to be certain that the method is not giving any false negative or positive signal.

1.2. Conformation-Sensitive Gel Electrophoresis

While SSCP looks at single stranded molecules, conformation-sensitive gel electrophoresis (CSGE) investigates conformation polymorphism of double stranded DNA molecules. During PCR, the complimentary strands of the amplified DNA molecules undergo repeated cycles of denaturation and renaturation. Heteroduplexes are formed in the presence of two distinct alleles of a DNA sequence. The generation of heteroduplexes containing looped out bases on one strand due to deletions or insertions were initially observed to give rise to aberrant electrophoretic migration when analyzed on regular non-denaturing polyacrylamide or agarose gels. The presence of such aberrantly migrating bands was first reported as PCR artifact *(9)*. Thereafter analysis of heteroduplexes by polyacrylamide gel electrophoresis became very common with best results coming from DNA loops or bubbles of three base pairs or larger. White et al. *(10)* had shown that heteroduplexes containing a single base mismatch within the looped portion of a hairpin structure have a different migration as compared to homoduplexes of the same DNA molecule.

The method of CSGE was developed for screening large fragments of DNA for single base differences such as SNPs and deletion as also insertions. In developing this method, the main emphasis was on the design of a method that is easy to use with standard laboratory equipment and reagents. Since the method is applicable to PCR products directly, it maximizes throughput and is highly efficient.

The hypothesis behind the development of CSGE was that an appropriate system of mildly denaturing solvents, like ethylene glycol and formamide, could amplify the conformational changes such as bends in the double helix produced by the presence of single-base mismatches. Thereby, the differential migration of DNA heteroduplexes and homoduplexes during gel electrophoresis can be enhanced.

The method of CSGE uses a non-denaturing polyacrylamide gel with the following modifications: (a) the crosslinker is 1,4-Bis acryloylpiperazine (BAP) instead of the traditional bis-acrylamide. This cross linker allows the use of a high concentration (10% final concentration) acrylamide gel for ease of handling but large pore size *(11)*. (b) a combination of two solvents, ethylene glycol and formamide, is used to enhance the structural perturbation of hetero-duplex molecules. (c) the gel buffer is based on Tris-taurine-EDTA, also known as glycerol tolerant gel buffer, to minimize distortions of the gel bands due to interaction of the borate ions with glycol based solvents like glycerol and others *(5)*.

The CSGE method has now been applied to analyses of a number of genes. All of these genes are large with multiple exons and many novel mutations and SNPs have been identified *(12–17)*.

2. Materials

2.1. Reagents

1. 40% Acrylamide solution (Gene Mate, CA), stored at 4°C (*see* **Note 1**).
2. 1,4-Bis acryloylpiperazine (BAP) (Fluka, NY), stored at 4°C.
3. Ethylene glycol (Sigma, MO), stored at room temperature.
4. Formamide (Gibco BRL, MD), stored at –20°C.
5. Ammonium persulphate (Amresco, OH), made fresh daily.
6. *N,N,N',N'*-tetramethylethylenediamine (TEMED) (Amresco), stored in the dark at 4°C (*see* **Note 1**).
7. 20X Gel buffer (TTE:Tris-Taurine-EDTA) (USB, OH), stored at room temperature.
8. Ethidium bromide solution (Sigma, MO), stored in the dark at room temperature (*see* **Note 2**).

2.2. Equipment

1. Standard manual DNA sequencing gel unit (BioMax STS-45I).
2. Glass plates 43 cm × 36 cm.
3. 0.8–1.0 mm-thick spacers for nonisotopic CSGE and 0.4-mm thick spacers for radioactive CSGE.
4. 36-well combs.
5. Power Supply units that can operate at constant voltage or constant wattage condition.
6. Large tray for holding CSGE gel during ethidium bromide staining.
7. Whatman 3MM paper.
8. UV-Transilluminator with large photodocumentation area.
9. Kodak MP4 polaroid camera.
10. Polaroid film type 667.
11. X-Ray autoradiograph films.

2.3. Solutions and Buffers

1. Running gel buffer (0.5X TTE): 44.5 mM Tris-14.25 mM Taurine-0.1 mM EDTA buffer, pH 9.0.
2. Stock solution of 20X TTE: 432 g Tris base, 144 g taurine, and 8.0 g disodium EDTA, were added to 2 L of water; pH usually comes to be 9.0 without any adjustment (*see* **Note 3**).
3. Stock solution of 40% acrylamide with 99:1 ratio of acrylamide:BAP: dissolve 404 mg of 1,4-bis acryloylpiperazine in 1 mL of water and add to 100 mL of 40% acrylamide solution (total volume, 101 mL).
4. Gel solution: 10% polyacrylamide (99:1 acrylamide: BAP; *see* below), 10% ethylene glycol, 15% formamide, 0.5X TTE.
5. Staining solution: 0.2 mL of ethidium bromide (10 mg/mL), 50 mL of 20X TTE, 1,950 mL sterile water (final concentration of 0.05 mg/mL ethidium bromide in 0.5X TTE).

3. Methods

3.1. Assembling the CSGE Gel Cassette

1. The glass plates for the gel, spacer, and the combs are cleaned every time before the assembly of the gel cassette. Hot water is the best cleaning agent followed by a rinse with deionized water (*see* **Note 4**).

2. The glass plates, spacers and comb are wiped clean with ethanol and dried with lint free tissue paper.

3. Silanize one of the glass plates to allow easy disassembly of the cassette at the end of the electrophoretic run.

3.2. Casting of the CSGE Gel Matrix

1. Prepare 175 mL of gel solution by mixing 43.75 mL of 40% 99:1(w/v) acrylamide:BAP solution, 4.4 mL of 20X TTE buffer, 26.25 mL of formamide, 17.5 mL of ethylene glycol, and 81.25 Millipore filtered water.

2. Filter and degas the acrylamide gel mixture for 20 min by vacuum filtration using a Nalgene 0.2μm (SFCA) bottle top filter attached to a glass bottle.

3. Add 1.75 mL of freshly prepared 10% ammonium persulfate solution and 100 μL of TEMED.

4. Pour gel immediately. Remove small bubbles formed during this process by slight tapping on the glass plates.

5. Allow the gel to polymerize for at least 2 h after casting the gel.

3.3. PCR Amplification and Heteroduplex Formation

Amplify target regions of genomic DNA using standard PCR protocol with the following modifications.

1. The use of a high fidelity Taq polymerase such as HiFidelity Taq Polymerase (Boehringer Mannheim, IN) is recommended to ensure elimination of errors due to incorporation of wrong bases during amplification.

2. To ensure heteroduplex formation, the last two cycles of PCR process are programmed to include incubation at 98°C for 5 min followed by slow cooling over 10 min to 68 °C and incubation at 68°C for 30 min (*see* **Notes 5–7**).

3.4. Gel Electrophoresis

1. Pre-Run gel at 750 volts for 15 min.

2. Wash sample wells with 0.5X TTE to ensure uniform stacking of the DNA samples on the gel surface (*see* **Notes 8,9**).

3. Mix 4–8 µL PCR product with 2 µL loading dye (30% formamide/ 0.25% bromophenol blue/0.25% xylene cyanol FF).
4. Load 4–5 µL of sample into each well (*see* **Note 10**).
5. Run samples at 400 volts for 16 h.
6. Monitor gel temperature by using thermometer strips (C.B.S. Scientific, CA). The adhesive-backed temperature indicator strip adheres easily to the inside of the glass plates and accurately monitors the gel temperature in the range of 35–75°C.

3.5. Staining

1. At the end of electrophoresis, disassemble the gel cassette so that the gel is left attached to one of the glass plates while the other plate is removed.
2. Stain the gel by layering just enough of the staining solution to cover the top surface of the gel for 5 min.
3. The gel surface should be perfectly horizontal such that the thin film of staining solution does not flow out.
4. After 5 min, destain the gel in distilled water for 10 min.

3.6. Photographic Documentation

1. Visualize the DNA bands initially with a handheld dual wavelength UV torch in the dark room (*see* **Note 11**).
2. Cut the relevant section of the gel with a scalpel and lift the gel section with a piece of dry Whatman 3MM blotting paper.
3. Release the gel section on the transilluminator by wetting the filter paper with water.
4. Photodocument the ethidium bromide stained bands under transillumination with an orange-red color correction filter using Polaroid type 667 film.

3.7. CSGE Using Radioactive PCR Products

The method of CSGE can also be adapted to using radioactive PCR products that remove the need for photography for gel documentation.

3.7.1. End Labeling of the PCR Primer with γ-P³³ ATP

1. Mix 1.6 µL of forward primer (10 pmol/µL), 2.0 µL 10X polynucle-otide kinase (PNK) buffer, 2.5 µL γ-P³³ ATP (10mCi/mL; 3000 Ci/mmol; New England Nuclear), 2.0 µL T4 PNK (1:10 dilution in PNK dilution buffer), and 11.9 µL water.
2. Incubate at 37°C for 30 min.
3. Inactivate the enzyme by heating the mixture for 10 min at 65°C.

3.7.2. PCR

1. Add 2.0 µL genomic DNA (20 ng) to a mixture containing 2.0 µL 10X PCR buffer, 3.2 µL dNTP (1.25 mM each), 1.0 µL 3'- primer (10 pmol/µL), 0.8 µL 5'- primer (10 pmol/µL), 2.5 µL labeled primer, 0.1 µL Taq polymerase, and 9.5 µL water.
2. Thermal cycle using the usual PCR conditions.

3.7.3. Gel Electrophoresis

1. Add 5.0 µL of loading buffer to each PCR tube.
2. Load 4.0 µL of each sample on gel.
3. Store the rest in the plastic beta blocking box in the freezer.
4. Due to higher sensitivity of autoradigraphy, a thinner CSGE gel matrix (0.4 mm) is used and separate the DNA species by gel electrophoresis at a high voltage (30 watts for 6 h).
5. After electrophoresis, the gel is dried on to Whatman 3MM filter paper and exposed to X-ray films for autoradiography.

4. Notes

1. Acrylamide and TEMED are potent neurotoxins. Gloves should always be worn when working with unpolymerized acrylamide solution.
2. Ethidium bromide is a carcinogen. Gloves should always be worn when working with ethidium bromide solution.
3. The stock solution of 20X TTE can be purchased as pre-made solution from USB, OH; Catalog No. 75827).
4. The presence of trace amounts of detergent on the glass plates can lead to smearing of the bands and can significantly disturb the reso-

lution. Therefore, it is imperative that the glass plates are washed meticulously and rinsed with hot water very carefully to remove any trace amounts of detergent.

5. When genomic DNA is used as a template for PCR, theoretically both alleles are amplified in equal proportion. During each cycle of PCR, amplified DNA molecules undergo denaturation and renaturation and generate homoduplex as well as heteroduplex molecules. However as the molar concentration of the amplified products increase in the later cycles, complete denaturation of products amplified in previous cycles may not happen. Therefore, some times it may be necessary to dilute the PCR products by a factor of 2 to ensure optimal denaturation and renaturation to favor heteroduplex formation.

6. One critical question about the CSGE technique is whether it will detect every SNP. The sequence context of a SNP clearly has an important effect on ease of detection by any physical, chemical or enzymatic method. As shown previously, a heteroduplex containing a C/T mismatch can be detected by differential migration of the heteroduplex if the C was in the sense strand but not if the C was in the antisense strand *(5)*. Many other observations suggest that as many as 5 nucleotides flanking a base mismatch may have an influence on the conformational change induced by the mismatch. Hence, it may be necessary to test as many as 4^{10} (or over a million) sequence contexts to ensure that a given technique can detect all possible mismatches. Thus, sequence context and nature of mismatch can modify degree of resolution. Under the standard conditions of CSGE as shown here, it has been shown that at least 90% of all SNPs are detected *(5,18)*. These CSGE conditions can always be modified to optimize the separation of any known SNP-bearing heteroduplex from the corresponding homoduplex molecules.

7. In modifying the CSGE conditions, the following factors may be considered. First, optimal resolution of heteroduplex from homoduplex molecules can be obtained for fragments 300–500 base pair in size. Second, centrally located mismatches are detected more easily than when located within 50 base pairs of either end of PCR product and can be missed *(5)*. Third, alteration of the concentration of the gel matrix from 8–15% can have remarkable effects on resolution of specific SNP-containing heteroduplex molecules.

8. The wells should be rinsed every time before loading sample to ensure that the starting front is very uniform — the degree of resolu-

tion depends to a large extent on shape of the starting sample front.

9. Multiplexing of up to four PCR products of different migration rates increases the throughput of the CSGE gels. Also individual PCR products can be pooled or loaded onto the same wells at definite time intervals (15 min is an optimum time difference). Routinely, $32 \times 3 = 96$ PCR products are loaded on a single CSGE run with four lanes available for markers. The resolution of the heteroduplex bands from the homoduplex bands remains the same whether loaded in batch or individually.

10. The sensitivity of detection can be a function of various factors induced by the experiment. For example, loading too little or too much DNA can mask the heteroduplex band. Theoretically for each denaturation/renaturation cycle, the amount of heteroduplex and homoduplex should be 50:50. In reality, the proportion of heteroduplexes detected range from 10% to 50%. This is determined by the amount of starting concentration of PCR products, GC-content of the DNA sequence, as well as relative migration of the two heteroduplex molecules with respect to the homoduplex molecules. It has been observed that for a particular SNP, complimentary heteroduplex molecules can have very different migration pattern with one of the two comigrating with the homoduplex molecules. In this scenario, the proportion of heteroduplex molecules observed will be the theoretical maximum of 25%. Furthermore, the number of distinct electrophoretically migrating homoduplex and heteroduplex bands can be as large as four representing two wild-type and mutant homoduplex molecules as well as two heteroduplex molecules. In contrast there can be just two bands where two homoduplex molecules comigrate as well as two heteroduplex molecules comigrate but distinct from the latter species.

11. UV light is damaging for the eyes and skin. Protective goggles and/or face shields, as well as gloves, should be worn always when working with UV light.

References

1. Gray, I. C., Campbell, D. A., and Spurr, N. K. (2000) Single nucleotide polymorphisms as tools in human genetics. *Human Mol. Genet.* **9**(16), 2403–2408.

2. Buetow, K. H., Edmonson, M. N., and Cassidy, A. B. (1999) Reliable identification of large numbers of candidate SNPs from public EST data. *Nat. Genet.* **21**(3), 323–325.

3. Shakked, Z. and Rabinovich, D. (1986) The effect of the base sequence on the fine structure of the DNA double helix. *Prog. Biophys. Mol. Biol.* **47**(3), 159–195.

4. Woodson, S. A. and Crothers, D. M. (1989) Conformation of a bulge-containing oligomer from a hot-spot sequence by NMR and energy minimization. *Biopolymers* **28**(6), 1149–1177.

5. Ganguly, A., Rock, M. J., and Prockop, D. J. (1993) Conformation-sensitive gel electrophoresis for rapid detection of single-base differences in double-stranded PCR products and DNA fragments: evidence for solvent-induced bends in DNA heteroduplexes. [erratum appears in *Proc. Natl. Acad. Sci. USA* 1994 May 24; **91**(11):5217]. *Proc. Natl. Acad. Sci. USA* **90**(21), 10,325–10,329.

6. Hayashi, S., Mori, I., Nonoyama, T., and Mitsumori, K. (1998) Point mutations of the c-H-ras gene in spontaneous liver tumors of transgenic mice carrying the human c-H-ras gene. *Toxicol. Pathol.* **26**(4), 556–561.

7. Gonen, D., Veenstra-VanderWeele, J., Yang, Z., Leventhal, B., and Cook, E. H., Jr. (1999) High throughput fluorescent CE-SSCP SNP genotyping. *Mol. Psychiatry* **4**(4), 339–343.

8. Oto, M., Miyake, S., and Yuasa, Y. (1993) Optimization of non-radioisotopic single strand conformation polymorphism analysis with a conventional minislab gel electrophoresis apparatus. *Anal. Biochem.* **213**(1), 19–22.

9. Nagamine, C. M., Chan, K., and Lau, Y. F. (1989) A PCR artifact: generation of heteroduplexes. *Am. J. Hum. Genet.* **45**, 337–339.

10. White, M. B., Carvalho, M., Derse, D., O'Brien, S. J., and Dean, M. (1992) Detecting single base substitutions as heteroduplex polymorphisms. *Genomics* **12**(2), 301–306.

11. Williams, C. J., Rock, M., and Considine, E., et al. (1995) Three new point mutations in type II procollagen (COL2A1) and identification of a fourth family with the COL2A1 Arg519—>Cys base substitution using conformation sensitive gel electrophoresis. *Human Mol. Genetics* **4**(2), 309–312.

12. Aradhya, S., Courtois, G., and Rajkovic, A., et al. (2001) Atypical forms of incontinentia pigmenti in male individuals result from mutations of a cytosine tract in exon 10 of NEMO (IKK-gamma). *Am. J. Human Genet.* **68**(3), 765–771.

13. Liu, W., Dong, X., and Mai, M. (2000) Mutations in AXIN2 cause colorectal cancer with defective mismatch repair by activating beta-catenin/TCF signalling. *Nat. Genet.* **26**(2), 146–147.

14. Has, C., Bruckner-Tuderman, L., and Muner, D., et al. (2000) The Conradi-Hunermann-Happle syndrome (CDPX2) and emopamil binding protein: novel mutations, and somatic and gonadal mosaicism. *Human Mol. Genet.* **9**(13), 1951–1955.

15. Melkoniemi, M., Brunner, H. G., and Manourrier, S., et al. (2000) Autosomal recessive disorder otospondylomegaepiphyseal dysplasia is associated with loss-of-function mutations in the COL11A2 gene. *Am. J. Human Genet.* **66**(2), 368–377.

16. Bignell, G. R., Warren, W., and Seal, S., et al. (2000) Identification of the familial cylindromatosis tumour- suppressor gene. *Nat. Genet.* **25**(2), 160–165.

17. Finnila, S., Hassmen, I. E., Ala-Kokko, L., and Majamaa, K. (2000) Phylogenetic network of the mtDNA haplogroup U in Northern Finland based on sequence analysis of the complete coding region by conformation-sensitive gel electrophoresis. *Am. J. Human Genet.* **66**(3), 1017–1026.

18. Korkko, J., Annunen, S., Pihlajamaa, T., Prockop, D.J., and Ala-Kokko, L. (1998) Conformation sensitive gel electrophoresis for simple and accurate detection of mutations: comparison with denaturing gradient gel electrophoresis and nucleotide sequencing. *Proc. Nat. Acad. Sci. USA* **95**(4), 1681–1685.

5

Detection of Mutations in DNA by Solid-Phase Chemical Cleavage Method

A Simplified Assay

Chinh T. Bui, Jeffrey J. Babon, Andreana Lambrinakos, and Richard G. H. Cotton

1. Introduction

Chemical Cleavage of Mismatch (CCM) is one of the methods of choice for mutation research and diagnosis of inherited diseases, as it is capable of detecting 100% of single-base mismatches *(1)*. The scientific background of CCM stems from the initial study of sequencing technique *(2)* in conjunction with other advanced studies associated with thermodynamics and secondary structures of single base pair mismatched DNA or RNA *(3)*. Such literature data confirmed that the mismatch point is locally destabilized and highly susceptible to many enzymatic *(4,5)* and chemical reactions *(6)*. Based on this platform, the CCM technology theoretically establishes the simplest chemical means to detect mismatch and thus mutation at this point in time. The method employs two commercially available chemicals, hydroxylamine *(7)* and potassium permanganate *(8-10)* to react with unmatched cytosine and thymine, respectively. The modification of the mismatch is then followed by

From: *Methods in Molecular Biology, vol. 212:*
Single Nucleotide Polymorphisms: Methods and Protocols
Edited by: P.-Y. Kwok © Humana Press Inc., Totowa, NJ

cleavage with piperidine and the resulting DNA fragments are simply analyzed by denaturing polyacrylamide gel-electrophoresis to identify the mismatch sites. Since the first protocol was described in 1988 *(11)*, the performance of this method has been continuously improved and the present protocol has some major advantages: (1) potassium permanganate ($KMnO_4$) has replaced the toxic osmium tetroxide (OsO_4) *(9,10)*; (2) if both mutant and wild-type DNA samples are labeled a double chance of mutation detection occurs; (3) the method is sensitive to as low as 0.1 µg of DNA samples; (4) and more importantly, all reaction steps are now being carried out on a silica bead solid support for convenient manipulation (*see* **Fig. 1**). Other alternative versions for mismatch detection have been established on the basis of enzymatic cleavage *(4,5)*. These methods are out of the scope of this protocol and they are briefly discussed for comparative purposes (*see* **Note 1**).

1.1. Strategy

The method involves the formation of heteroduplex DNA of two complementary types (*see* **Fig. 2**) which is generated by the melting and re-annealing of the mutant and the wild-type (control) DNA. If the two sequences (mutant and wild-type) are different at any oligonucleotide base, a complementary pair of single base pair mismatches will be generated and mismatched C and T bases will be susceptible to chemical modification and cleavage. Because all classes of C and T mismatches (CC, CT, CA, TT, TG, and TC) are cleaved, a complete screening for point mutation can be achieved using only wild-type (or mutant) DNA as probe (*see* **Fig. 2**). However, to obtain two chances of detecting a mutation, labeling of both wild-type and mutant DNA is recommended (*see* **Note 2** and **Fig. 2**). The DNA probes can be synthetic oligonucleotides or PCR amplification products of specific genomic DNA sequences.

2. Materials

1. TE buffer: Mix 100 µL of 1 *M* Tris-HCl, pH 8.0, 20 µL of 0.5 *M* ethylenediamine-tetraacetic acid (EDTA, Aldrich) and 9.88 mL of distilled water. Store the TE buffer at room temperature.

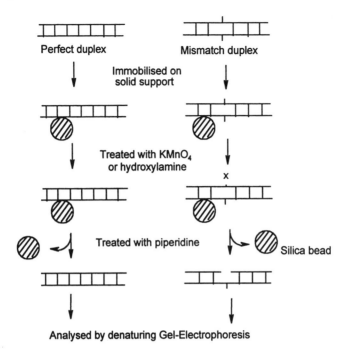

Fig. 1. Solid-phase chemical cleavage of mismatch. Both perfect and mismatch duplexes are immobilized on silica beads. Chemical modification reactions are carried out while DNA duplexes still remain on solid support. One sample of DNA is treated with hydroxylamine and another with potassium permanganate. Piperidine treatment simultaneously cleaves the mismatched point and releases the samples for gel electrophoretic analysis.

2. 4.2 M Hydroxylamine solution: 1.39 g solid hydroxylamine hydrochloride (Aldrich) is dissolved in 1.6 mL distilled water. The resulting solution is adjusted with diethylamine (Aldrich) to pH = 6.0 (ca. 1 mL of diethylamine is required). Water is added to adjust the final volume to 4 mL. Store at –20°C for up to 6 mo.

3. 3 M tetraethylammonium chloride (TEAC) solution: 49.7 g of tetraethylammonium chloride (Aldrich) is dissolved in 100 mL distilled water. Store at 4°C for up to 3 mo.

4. 1 mM KMnO$_4$ solution: 80 mg of KMnO$_4$ (Aldrich) is dissolved in 5 mL of distilled water. 10 µL of the resulting solution is mixed by vortex with 900 µL of 3 M TEAC solution to give 1 mM KMnO$_4$ solution. Prepare the solution freshly before use.

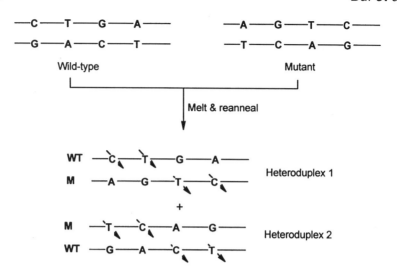

Fig. 2. All mutations have two chances of being detected by use of labeled DNA of both senses of mutant (M) and wild-type (WT) DNA in making the heteroduplexes. Arrows represent cleavage at mismatched T and C bases.

5. Cleavage-dye solution: Add 20 µL undiluted piperidine (Aldrich) to 64 µL formamide (Amresco, USA) and 16 µL dye (50 mg blue dextran [Aldrich] per mL). Store at 4°C for 1 d.

6. Gene amplification PCR kit is commercially available from Perkin Elmer (Foster City, CA). The AmpliTaq Gold™ contains GeneAmp buffer, MgCl₂ solution and AmpliTaq DNA polymerase.

7. Fluorophore 6-FAM and HEX for the 5' and 3' primers, respectively, can be purchased from Geneset Corp. (La Jolla, CA).

8. Solid support for DNA is commercially available from Mo Bio Laboratories Inc. The UltraClean™ DNA purification kit contains: Ultra-salt solution, Ultra-bind bead, and Ultra-wash solution, which are used in this protocol.

9. PCR purification step: the DNA is separated from PCR primers, unincorporated nucleotides, buffers, and enzyme by using Strata™ PCR Purification Kit. The kit can be obtained from Stratagene (La Jolla, CA) (*see* **Note 3**).

10. Tris-borate EDTA (TBE) buffer for electrophoresis: Mix 16.2 g Tris-base, 8.1 g boric acid, and 1.12 g EDTA in 1500 mL distilled water, pH = 8.0. Store at 25°C.

3. Methods

The assay should be carried out in a fume hood as hydroxylamine, acrylamide, formamide, and piperidine are noxious chemicals.

3.1. DNA Preparation

1. Amplify plasmid DNA (about 0.1 µg) using a GeneAmp PCR System 9700 (PE Biosystems) with fluorescent-labeled primers (6-FAM for the 5' primer, HEX for the 3' primer) (*see* **Note 2**).
2. Purify the resulting DNA samples (wild-type and mutant) by using the purification kit (Stratagene; *see* **Note 3**) or cutting a band from an agarose gel.
3. Determine the concentrations of DNA samples by either measuring the absorbancy at 260 nm (UV-visible spectrophotometer) or using the molecular weight standards for agarose gel electrophoresis (DNA ladder, New England BioLabs Inc.).

3.2. Formation of Heteroduplex DNA

1. Mix by vortex equal amounts of the labeled wild-type and mutant DNA in TE buffer.
2. Apply the mixture to the PCR machine for heteroduplex formation by heating the sample to 99°C for 7 min, then cooling it down to 65°C and maintaining at this temperature for 1 h.
3. Finally, cool the DNA sample to 25°C for 30 min (*see* **Note 4**).

3.3. Attachment of Homo- and Heteroduplexes DNA onto Silica Beads

1. Place 1 µL of heteroduplex DNA samples (0.1–0.2 µg DNA) into 2 separate Eppendorf tubes (one is labeled for Het/KMnO$_4$ assay and one for Het/hydroxylamine assay).
2. Place 1 µL of homoduplex DNA samples (0.1–0.2 µg DNA) into 2 separate Eppendorf tubes (one is labeled for homo/KMnO$_4$ assay and one for homo/hydroxylamine assay).
3. Add 2.5 µL of Ultra-bind bead suspension to all four Eppendorf tubes (*see* **Notes 5,6**).
4. Gently mix the tubes on shaker at room temperature for 1–2 h.

5. Wash the DNA bound beads carefully with Ultra-wash solution (2 × 200 μL).
6. Dry the beads in open air at 25°C for 15 min.

3.4. Reaction with KMnO₄

1. Add 30 μL of 1 m*M* KMnO₄ in 3 *M* TEAC solution (*see* **Notes 7,8**) into two Eppendorf tubes that are labeled with Het/KMnO₄ assay and Homo/KMnO₄ assay (*see* **Note 8**).
2. Incubate the tubes at 25°C for 10 min (*see* **Note 9**).
3. Centrifuge the tubes at 325*g* and carefully decant the supernatant (by Pasteur pipet).
4. Wash the pellet twice with 200 μL Ultra-wash solution.
5. Dry the beads in open air for 15 min.

3.5. Reaction with Hydroxylamine

1. Add 30 μL of 4.2 *M* hydroxylamine in TEAC solution to two Eppendorf tubes labeled with Het/hydroxylamine assay and Homo/hydroxylamine assay.
2. Incubate the reaction mixture at 37°C for 40 min (*see* **Note 9**).
3. Centrifuge the tube and carefully decant the supernatant.
4. Wash the pellet twice with 200 μL of Ultra-wash solution.
5. Dry the beads in open air at 25°C for 15 min.

3.6. Cleavage by Piperidine

1. Add 10 μL of cleavage dye solution to all four reaction tubes.
2. Heat the reaction tubes to 90°C and maintain at this temperature for 30 min.
3. Cool the tubes on ice and the solid beads are separated by centrifuge (*see* **Note 10**).
4. Load the supernatant on to a denaturing polyacrylamide gel 4.25% (19:1) acrylamide:*bis*-acrylamide, 6 *M* urea gel, run by ABI 377 DNA sequencer using TBE buffer at 3000V. Electrophoresis will take approx 3 h for analysis of a 500 bp fragment.

3.7. Result Analysis

Mismatch detection is based on comparative study between traces for homoduplex and heteroduplex DNA samples. Cleavage peaks present in the trace of heteroduplex sample but not the control (homo) represent the mutation. (For typical example, *see* **Figs. 3 and 4** and **Note 11**.)

4. Notes

1. Based on the reactivity of mismatched bases in RNA and DNA three methods have emerged: a nonradioactive cleavage method by ribonuclease A *(5)*, the enzymatic mismatch cleavage (EMC) *(4)*, and the chemical mismatch cleavage (CCM). The EMC is commercially available as a Passport EMD™ kit (Amersham Pty Ltd.) and relies on the ability of an enzyme, T4 endonuclease VII, to cleave base-pair mismatches. The major advantages of the EMC are: (1) the enzyme binds and cleaves at the mismatched point in a one-step reaction while the CCM requires two-step process; (2) one enzyme can recognize all types of mismatches, whereas the CCM requires two types of chemicals to achieve the same purpose. However, the EMC is more expensive and often suffers from the need to optimize thoroughly the reaction conditions, and substantial cleavage of matched bases occurs leading to high background bands. The ribonuclease method needs the production of RNA to form the duplexes, but as the cleavage at the mismatches is double-stranded it can be analyzed on an agarose gel. A kit is available from Ambion (USA).

2. False-positives and -negatives have not been reported so far. However, labeling both mutant and wild-type DNA will offer two chances of mutation detection in the event of the rare occurrence of unreactive mismatch. Radioactive labels can also be used instead of fluorescent ones.

3. PCR products of mutant and wildtype DNA can be conveniently purified by using the Stratagene purification kit or by agarose gel-electrophoresis. In the latter case, the band is precisely cut and loaded on to silica beads.

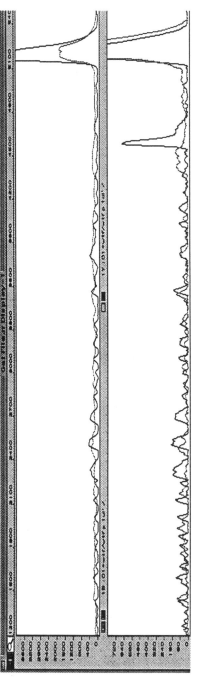

Fig. 3. Detection of a TC mismatch in a 540 bp DNA fragment by using KMnO$_4$ / piperidine assay. The control trace (homoduplex, top) shows no cleavage peak and the mismatched DNA trace (heteroduplex, bottom) displays a strong cleavage peak of the mismatch T base in the 5' FAM sequence. Note consistent background in both traces, which is essentially a chemical sequencing trace of T bases allowing confidence that reaction has occurred and a position reference.

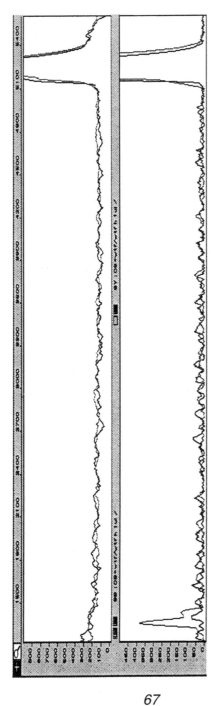

Fig. 4. Detection of TC mismatch of 540 bp DNA fragment by using hydroxylamine/piperidine assay. The control trace (homoduplex, top) shows no cleavage peak and the mismatched DNA trace (heteroduplex, bottom) displays a strong cleavage peak of 3'HEX sequence at the mismatched C base. Note this is the second chance of detecting the mutation, the first being in **Fig 3**.

4. Size, concentration, and temperature of the melting heteroduplex formation should be taken into account as there can be poor heteroduplex formation in some cases. For example, for certain heteroduplex
   formation with GC-rich DNA, boiling to 100°C should be used. A
   test should be carried out on agarose gel to make sure that no intense
   multiple bands or smeary bands are present after heteroduplex
   formation.

5. Attachment of homo- and heteroduplex DNA onto the commercially
   available silica beads are the first important step in solid-phase CCM
   method. The adsorption is achieved under relatively high concentration (3 *M*) of TEAC salt solution and the DNA molecule remains
   attached throughout the modification and washing steps.

6. The DNA length is limited by the analytical technique, fidelity of the
   heteroduplex formation, and the solid supports. In our study, the
   silica beads and special conditions are most suitable for up to 500 bp
   fragments of DNA. Refer to liquid-phase protocol for larger fragments (1–2 kb) *(12)*.

7. In principle, bases other than TEAC (e.g., Tetramethylammonium
   chloride, TMAC) can be used in this assay. However, TEAC is superior to others in terms of less toxicity and high reactivity of DNA
   structure. In this assay, TEAC salt functions as a destabilizer of
   duplex DNA helix and therefore it increases the reactivity of nucleotide bases with hydroxylamine and $KMnO_4$.

8. Aqueous $KMnO_4$ solution should be freshly made before use. The
   aging solution (after 1 d) turns brown-yellow with precipitation of
   MnO_2. The reaction is dependent on temperature and concentration
   of substrates. Usually the concentration of chemical given is correct
   for approx 100–200 ng of total weight of DNA.

9. Prolonged incubation can lead to overreaction and destruction of the
   heteroduplexes and fragment DNA. Underincubation can give rise
   to no cleavage bands. Time courses of incubation are recommended
   when starting to use this test.

10. Both mismatch cleavage and release of DNA from silica beads are
    achieved in one-step reaction. Separation of excess piperidine prior
    to the gel electrophoresis step is not required.

11. In one typical example, the CCM method was applied to detection of
    a TC mismatch in a 540 bp DNA fragment derived from the cloned
    mouse β-globin promoter DNA. Both mutant and wild-type DNA
    samples were amplified using fluorescent-labeled primers (6-FAM

for the 5' primer, HEX for the 3' primer). Formation of DNA homo- and heteroduplexes were performed under the standard conditions and subjected to the solid-phase CCM procedure as described earlier. The results of the cleavages are shown after electrophoresis and analysis on an ABI-377 sequencer. **Figures 3** and **4** show single and strong cleavage peaks as the result of cleavage reaction of T and C mismatches by $KMnO_4$ (*see* **Fig. 3**) and hydroxylamine (*see* **Fig. 4**), respectively, as compared to the control without any cleavage peak detected.

References

1. Ellis, T. P., Humphrey, K. E., Smith, M. J., and Cotton, R. G. H. (1998) Chemical cleavage of mismatch: a new look at an established method/recent developments. *Human Mutat.* **11**, 345–353.

2. Maxam, A. M. and Gilbert W. (1977) A new method for sequencing DNA. *Proc. Natl. Acad. Sci. (USA)* **74**, 560–564.

3. Kennard, O. (1988) Structural studies of base pair mismatches and their relevance to theories of mismatch formation and repair, in *Structure and Expression*, vol. 2, (Sarma, R. H. and Sarma, M. H., eds.), Academic Press, New York, NY, pp. 1–25.

4. Youil, R., Kemper, B. W., and Cotton, R. G. H. (1993) Screening for mutations by enzyme mismatch cleavage using T4 endonuclease VII. *Am. J. Hum. Genet.* **53**, Abstract 1257.

5. Myers, R. M., Larin, Z., and Maniatis, T. (1985) Detection of single base substitution by ribonuclease cleavage at mismatches in RNA:DNA duplexes. *Science* **230**, 1242–1246.

6. Smooker, P. M., and Cotton, R. G. H. (1993) The use of chemical reagents in detection of DNA mutations. *Mutation Res.* **288**, 65–77.

7. Cotton, R. G. H. (1989) Detection of single base changes in nucleic acids. *J. Biochem.* **253**, 1–10.

8. Gogos, J. A., Karayiorgou, M., Aburatani, H., and Kafatos, F. C. (1990) Detection of single base mismatches of thymine and cytosine residues by potassium permanganate and hydroxylamine in the present of tetralkylammonium salts. *Nucleic Acids Res.* **18**, 6807–6812.

9. Roberts, E., Deeble, V. J., Woods, C. G., and Taylor, G. R. (1997) Potassium permanganate and tetraethylammonium chloride are a safe and effective substitute for osmium tetroxide in solid-phase fluorescent chemical cleavage of mismatch. *Nucleic Acids Res.* **25**, 3377–3378.

10. Lambrinakos, A., Humphrey, K. E., Babon, J. J., Ellis, T. P., and Cotton, R. G. H. (1999) Reactivity of potassium permanganate and tetraethylammonium chloride with mismatched bases and a simple mutation detection protocol. *Nucleic Acids Res.* **27**, 1866–1874.

11. Cotton, R. G. H., Rodrigues, H. R., and Campbell, R. D. (1988) Reactivity of cytosine and thymine in single base-pair mismatches with hydroxylamine and osmium tetroxideand its application to the study of mutations. *Proc. Natl. Acad. Sci. USA* **85**, 4397–4401.

12. Cotton, R. G. H. (1999) Detection of mutations in DNA and RNA by chemical cleavage, in *The Nucleic Acid Protocols Handbook* (Methods in Molecular Biology Series, Rapley, R., ed.), Humana Press, Totowa, NJ, pp. 685–693.

6

SNP Discovery by Direct DNA Sequencing

Pui-Yan Kwok and Shenghui Duan

1. Introduction

DNA sequencing, while relatively laborious, is the gold standard in mutation detection and single nucleotide polymorphism (SNP) discovery. The most widely used approach is direct DNA sequencing of polymerase chain reaction (PCR) products with dye-terminator chemistry analyzed on automated DNA sequencers *(1)*. Although the quality of DNA sequencing data has improved significantly over the last few years, the peak pattern remains uneven and random artifacts are seen from time to time *(2)*. Because human cells are diploid, DNA sequence of a heterozygote contains a locus where two different bases occupy the same site. The uneven peak pattern makes it difficult sometimes to discern these composite peaks because one of the two polymorphic bases may be disproportionately smaller than the other base and the base-calling algorithm of the automatic DNA sequencer misses the correct call *(2–4)*.

Fortunately, the peak pattern of a DNA sequence is highly reproducible and is determined by the local sequence context *(2–5)*. In other words, if the same segment of DNA is amplified from a number of different individuals and the resultant PCR products are

From: *Methods in Molecular Biology, vol. 212:*
Single Nucleotide Polymorphisms: Methods and Protocols
Edited by: P-Y. Kwok © Humana Press Inc., Totowa, NJ

sequenced, all the samples will yield the same peak pattern, regardless of the origins of the DNA sample or when the DNA sequencing is done. Therefore, when the DNA sequencing traces of multiple individuals are compared to each other, the peak patterns of the heterozygotes and the homozygotes are noticeably different and the mutations or polymorphisms can be identified easily *(5)*.

In addition, the relative peak heights of the polymorphic bases can be used to estimate allele frequencies when pooled DNA samples are sequenced and compared to a reference DNA sequence. While the pooled DNA sequencing approach does not have the necessary resolution to distinguish pooled samples with small allele frequency differences, significant discrepancies ($\geq 10\%$) are easily identified *(1)*.

In this chapter, in addition to describing the DNA sequencing of purified PCR products to identify SNPs, we also describe a streamlined DNA sequencing approach that does not require post-PCR processing prior to DNA sequencing and the analysis algorithm used for allele frequency estimation.

The key to robust SNP detection by DNA sequencing is high-quality sequencing data. Accordingly, PCR primer design that emphasizes specificity and high yield is of great importance. A good primer design program is the modified Primer3 *(6,7)*. If the PCR primers selected are found to be unique in the genome by homology searches against the human genome DNA sequence, the chances of obtaining a specific PCR product are high. Even if the PCR is not specific, one can obtain good sequencing data by purifying the PCR mixture to obtain the desired PCR product.

When the desired PCR product is the only species generated in the PCR reaction, one can use the product directly in the sequencing reaction without purification. The protocol is further simplified by using an asymmetric PCR approach, performing the amplification with a 10:1 mixture of the PCR primers and a reduced amount of deoxyribonucleotide triphosphates (dNTPs). In this configuration, the PCR primer at the lower concentration is used up during PCR, leaving the excess PCR primer as the sequencing primer in the next step. The excess dNTPs do not interfere with the sequencing reaction because the sequencing mix contains a much higher concentration of dNTPs.

At the end of the sequencing reaction, the dye-terminators are removed by size-exclusion chromatography using spin columns. Spin-column purification produces much better quality data than those generated by ethanol precipitation. The part of the sequencing trace with the highest quality is the 400 bp segment between 50–450 bases from the 3'-end of the sequencing primer. The sequencing data can be made more uniform from sample to sample by "trimming" the low-quality data at the beginning and end of the sequencing traces before reanalyzing the data. If done the same way for all samples for the same marker, the sequencing traces can be compared more easily. By comparing the peak patterns of sequencing traces from a number of individuals, one can identify differences between them at the polymorphic sites.

When homozygotes of different alleles are present among the samples sequenced, the polymorphisms can be identified easily using any sequence alignment programs. In cases where the minor allele frequency is low, one usually sees only homozygotes of one allele and a handful of heterozygotes. Here, one relies on a break in the peak pattern where the heterozygous samples exhibit a peak whose height is reduced by half (when compared to the homozygotes) together with the telltale sign of a second base underneath and an often observed phenomenon of a change in peak height in the base 3'- to the polymorphic base *(5)*.

If equal amounts of genomic DNA from a group of individuals are pooled together, the pooled samples can be amplified and sequenced as usual. The pooled DNA sequencing trace can be compared against a reference sequencing trace for allele frequency estimation. Although the resolution of the estimates is not perfect, 10% differences between two pools can be detected with confidence.

2. Materials

2.1. Reagents

2.1.1. PCR

1. Thermostable DNA Polymerase AmpliTaq Gold™ at 5U/µL (Applied Biosystems, Foster City, CA).
2. 10X PCR Buffer II (Applied Biosystems).
3. 25 mM MgCl$_2$ solution (Applied Biosystems).
4. dNTP mixture: 2.5 mM dATP, 2.5 mM dCTP, 2.5 mM dGTP, and 2.5 mM dTTP.
5. PCR primers are designed by modified Primer3 program *(6,7)*.
6. Skirted 96-well white PCR plate (Marsh Bio Products, Rochester, NY).
7. Easy-peel heat-sealing foil (Marsh Bio Products).
8. Silicone compression mats (Marsh Bio Products).
9. Strip tubes (200 µL) (Midwest Scientific, Valley Park, MO).
10. Strip caps (Midwest Scientific).
11. MicroAmp Optical 96-well reaction plates (Applied Biosystems).

2.1.2. Sequencing

1. ABI Prism®BigDye™ Terminator v3.0 (Applied Biosystems).
2. 5X Sequencing buffer (Applied Biosystems).
3. Thermowell sealer (aluminum) (Corning Incorporated, Corning, NY).

2.1.3. PCR and Sequencing Product Purification

1. Low-melting-point agarose (UltraPure) (Life Technologies, Rockville, MD).
2. 1X Tris-acetate-EDTA (TAE): 0.04 M Tris-acetate, 0.001 M EDTA, pH 8.0.
3. Wizard® PCR Preps DNA Purification System (Promega, Madison, WI).
4. 6X Loading buffer: 0.25% (w/v) bromophenol blue, 0.25% (w/v) xylene cyanol FF, 30% (v/v) glycerol in water.
5. Ethidium bromide (10 mg/mL).
6. Centri-Sep 8 Strips (Princeton Separations, Adelphia, NJ).
7. Centri-Sep 96 (96-well gel filtration plate, Princeton Separations).

2.2. Equipment

1. Thermocycler.
2. Horizontal gel electrophoresis devices.
3. Thermo-sealer (Marsh Bio Products).
4. ABI Prism® 3700 DNA Analyzer with DNA Sequencing Analysis Software™. Version 3.6.1 (Applied Biosystems).

3. Methods

3.1. DNA Sequencing with Purified PCR Products

3.1.1. PCR Reaction

1. Amplify genomic DNA in 30 μL reaction mixtures containing 3 μL of genomic DNA (12 ng), 3.0 μL of 10X PCR buffer II, 4.2 μL of 25 mM MgCl$_2$, 2.4 μL of 2.5 mM dNTP mixture, 6 μL of 1 μM each PCR primer, 0.15 μL (0.75 U) of AmpliTaq Gold DNA polymerase, and 11.25 μL ddH$_2$O.
2. Activate the AmpliTaq Gold DNA polymerase by heating the reaction mixture at 95°C for 12 min. Perform PCR using 35 cycles of denaturation at 92°C for 10 s, primer annealing at 58°C for 20 s, and primer extension at 68°C for 30 s.
3. Incubate the reaction mixture at 68°C for 10 min for final primer extension and hold it at 4°C until further use.

3.1.2. Purification of PCR Products from Low-Melting-Point Agarose Gel

1. Prepare 0.8% (w/v) low-melting-point agarose gel with 1X TAE, with 3 μL of 10 mg/mL ethidium bromide added to each 100 mL of gel solution. Run electrophoresis in 1X TAE buffer *(5)*.
2. Add 6 μL of 6X loading buffer to the PCR mixture (30 μL). Load the entire 36 μL reaction mixture onto the agarose gel.
3. Perform electrophoresis at 4 V/cm; running time will depend on the product size.
4. Excise the desired DNA-containing gel slices under long UV (365 nm) transillumination (*see* **Notes 1** and **2**).
5. Transfer the gel slices to 1.5-mL microcentrifuge tubes, purified by using the Wizard PCR Preps DNA Purification System kit according to the protocol provided with the kit.

6. Incubate tubes containing gel slices in 70°C water bath until gel is completely melted. Add 1 mL resin and mix thoroughly for 20 s. (Do not vortex!)

7. For each sample prepare one Wizard minicolumn attached to a syringe barrel and insert it to the vacuum manifold, add the DNA-resin mixture to the syringe barrel, and apply vacuum until all liquid passes through minicolumn.

8. Wash minicolumn by adding 2 mL of 80% isopropanol to the syringe barrel and applying vacuum to pull solution through minicolumn. Air dry resin by applying vacuum for an additional 30 s.

9. Remove the syringe barrel and centrifuge the minicolumn at 10,000g for 2 min in a 1.5-mL microfuge tube. Discard the washing.

10. Transfer the minicolumn to a clean 1.5-mL microfuge tube. Elute DNA by adding 50 µL ddH$_2$O to the minicolumn and incubate at room temperature for 1 min; follow this with centrifugation at 10,000g for 20 s.

11. Store purified PCR product at 4°C until further use.

3.1.3. Sequencing Reaction

1. Assemble the sequencing reaction (12 µL total volume) by adding 5 µL of purified PCR product to a strip tube containing 2.0 µL of BigDye™ Terminator v3.0, 1 µL of 5X sequencing buffer, 1 µL of sequencing primer (2 µM), and 3 µL of ddH$_2$O (*see* **Note 3**).

2. Denature the DNA initially by incubating the reaction mixture at 96°C for 2 min.

3. Perform cycle sequencing with 26 cycles of denaturation at 96°C for 15 s, primer annealing at 50°C for 1 s and primer extension at 60°C for 4 min.

4. Hold the product mixture at 4°C until further use.

3.2. DNA Sequencing with Crude PCR Products

3.2.1. Asymmetric PCR

1. Amplify genomic DNA by adding 1 µL of DNA (4 ng) to a mixture containing 1 µL of 10X PCR buffer II, 1.4 µL MgCl$_2$ (25 mM), 0.4 µL of 2.5 mM dNTP mixture, 5 µL of PCR primers mixture (2 µM for one primer and 0.2 µM for the second primer), 0.03 µL (0.15 U) of AmpliTaq Gold DNA polymerase, and 1.17 µL ddH$_2$O. Total

reaction volume is 10 μL. Thermocycling conditions are the same as those found in **Subheading 3.1.1.** (*see* **Note 4**).

3.2.2. Sequencing Reaction

1. For sequencing add 2.5 μL of the crude asymmetric PCR product to a mixture containing 2 μL of BigDye™ Terminator v3.0, 1 μL of 5X sequencing buffer, and 6.5 μL ddH$_2$O. Thermocycling conditions are the same as those found in **Subheading 3.1.3.**

3.3. Purification of Sequencing Products

1. Add 12 μL of ddH$_2$O to the 12 μL of sequencing reaction products (*see* **Note 5**).
2. For a small number of samples, bring the desired number of Centri-Sep 8 strips to room temperature before use. Remove the top foil and integral bottom of the strips and spin for 2 min at 750*g* to remove the storage liquid.
3. Transfer the samples onto the center of the gel bed without disturbing the gel surface and place the strips onto clean PCR strip tubes.
4. Collect the samples by centrifugation for 2 min at 750*g*. Hold samples at 4°C and placed on the capillary DNA sequencer for analysis.
5. For a large number of samples, use the Centri-Sep 96-Well gel filtration plates purification approach. Bring plates to room temperature and remove adhesive foils from the bottom and then the top.
6. Place the plate on top of a 96-Well wash plate and centrifuge at 1500*g* for 2 min to remove the storage liquid.
7. Transfer sequencing reaction mixtures onto individual wells of the plate, taking care that the samples are loaded onto the center of the gel bed without disturbing gel surface.
8. Place the gel plate on top of a clean 96-Well collection plate and centrifuge at 1500*g* for 2 min. Hold the purified sequencing products at 4°C until loading onto the capillary DNA sequencer (*see* **Note 6**).

3.4. SNP Identification

Sequence at least two different individuals for sequence comparison. After the sequencing run on ABI Prism® 3700 DNA analyzer, trim and reanalyze sequencing traces with the DNA Sequencing

Analysis Software™, Version 3.6.1. The goal is to remove poor quality bases from the sequencing trace and boost sequencing signal the same way for all the sequences being compared. This is done by modifying the "start point" and "end point" of the individual sequence traces and reanalyze them. The reanalyzed traces are assembled by Sequencher™ for Windows, Version 4.0.5 (Gene Code Corporation, Ann Arbor, MI). A Macintosh version is also available from the same company. Compare the peak patterns to each other by looking at the traces and positions flagged by Sequencher. When both alleles of a variation are found in homozygotes in the samples sequenced, the Sequencher program will designate the base position as being occupied by an N. If only one allele is represented by homozygotes while the second allele is only found in heterozygotes, it is less obvious to the computer program and one has to examine every base carefully. For instance, if only G/A heterozygotes and G/G homozygotes are found in the traces sequenced, and if the G peak is higher than the A peak in the heterozygous trace, the computer software will often call the composite G/A peak in the heterozygote as a G. However, in the sequencing trace with the heterozygote, the G peak height at the candidate SNP site is about half the size as the homozygous G peak, and it is always accompanied by a second peak underneath and an observable change in peak height in the base 3'- to the polymorphic base.

3.5. Allele Frequency Estimation

Software packages used are the same with the ones listed in **Subheading 3.4.**

1. Pool DNA samples by mixing equal amounts of DNA from any number of individuals. Typically, we prepare pools consisting of 5–100 individuals.
2. Dilute the DNA to a final concentration of 4 ng/µL. A reference DNA sample from one individual (4 ng/µL) is used as reference. For optimal results, we design PCR primers to place the SNP in the middle of the sequencing fragment or at least 100 bp away from the 3'-end of the primer to be used in sequencing.

3. Trim and reanalyze sequencing traces as described in **Subheading 3.4.** and use the Sequencher program for allele frequency estimation. Align the sequencing traces of the pool and reference samples.
4. Use the "Find" function of the program to identify the polymorphic site by entering a few bases of the flanking DNA sequence. With the chromatograms of all the traces opened on the same window, look at the reference trace first to determine the allele to use as reference.
5. If the reference sequence is from a homozygote, the base at the polymorphic site is the reference allele by default.
6. If the reference sample is a heterozygote, pick the base with the greater peak height as the reference allele.
7. Measure the peak heights of the reference allele and normalizing peak for the reference sample and the pooled samples (*see* **Note 7**).

To estimate the allele frequencies in the DNA pools, we normalize the sequencing traces of the pooled DNA samples and the reference sample to account for the global signal intensity differences from trace to trace. The normalization process is done by identifying a base of the same type and of similar height to the reference allele in the reference sample. For example, if the alleles of a SNP are C and T, and the reference sample is a heterozygote with the C peak being taller than the T peak, we find a C peak of similar height from 20-base windows upstream or downstream from the polymorphic site to serve as the normalizing base (*see* **Note 8**). We exclude the three bases immediately upstream or downstream from the polymorphic site because the height of these base peaks are sometimes influenced by the bases found in the polymorphic site. The normalizing base is then identified in the sequencing traces from the pools. The peak heights of the reference allele (C in our example) and the normalizing base (another C with the vicinity) for the reference and the pools are measured and the allele frequency is estimated as follows.

First, compute the ratio of the normalizing peak heights to obtain the normalizing factor

$$f = N_{ref}/N_{pool}, \tag{1}$$

where N_{ref} is the height of normalizing peak in the reference sequencing trace and N_{pool} is the height of normalizing peak in the pool sequencing trace.

Then, compute the allele frequency of the reference allele in the pool by taking the ratio of the peak heights of the polymorphic alleles using the following expression:

$$\text{Allele frequency of the selected allele in pool sample} = cf\,(P_{pool}/P_{ref}), \qquad (2)$$

where P_{pool} is the peak height of selected allele (of the same type as the reference peak, C in this example) in the pool sequence and P_{ref} is the peak height of the reference allele, and c is a constant to adjust for the status of the reference sample (0.5 when the reference sample is a heterozygote and 1.0 when the reference sample is a homozygote). Because f equals N_{ref}/N_{pool}, it is simpler to rearrange the expression when multiple pools are being estimated to:

$$\text{Allele frequency of the selected allele in pool sample}$$
$$= c(N_{ref}/N_{pool})(P_{pool}/P_{ref}) \qquad (3)$$
$$= c(P_{pool}/N_{pool})/(P_{ref}/N_{ref})$$

This way, one always takes the ratio of the peak height of the allele being estimated and the height of the normalizing peak from the same sequencing trace and computes the allele frequency by comparing the ratio found in the pool versus the ratio found in the reference, modified by the constant c depending on whether the reference sequence is from a heterozygote or a homozygote (see **Figs. 1** and **2** and **Note 9**).

The estimated allele frequency of the second allele is estimated by simply subtracting the allele frequency of the reference allele from 1 (see **Note 10**).

4. Notes

1. In order to avoid DNA degradation, do not excise gel band under UV light with shorter wavelengths.
2. To get better DNA recovery from low melting-point agarose gel, keep gel slice volume to less than 300 µL.

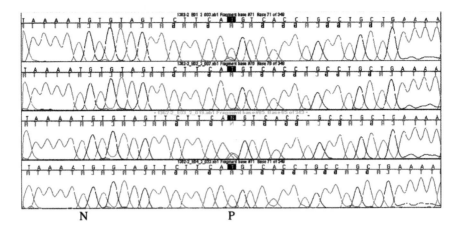

N   P

Fig. 1. Allele frequency estimation of SNP in 3 pooled samples. The sequencing traces are from top to bottom: African American, Japanese/Chinese, Caucasian, reference individual. The alleles are T and C, with the reference individual homozygous for T. The T allele is therefore the reference allele (with peak height P). The normalizing peak is the one most similar in height as that for the reference allele in the reference individual (with peak height N). By comparing the ratios of P/N of the pool sample and the reference sample, the allele frequencies are estimated as 81, 100, and 33% T, respectively, in the African American, Japanese/Chinese, and Caucasian panels. The allele frequencies for C are therefore 19, 0, and 67%, respectively.

3. The major cause of PCR failure when performing 10 µL reaction is that evaporation during PCR (to occupy the volume above the reaction mixture in the PCR tube) reduces the effective reaction volume and alters the reagent concentration. To solve this problem and work more efficiently, we add 2 µL of water to the reaction so that upon heating during PCR, the effective volume is back to 10 µL and the reaction works properly. In addition, we use thermo-sealer to seal the plates with easy peel heat sealing foils. A silicone compression mat is put on top of the plate when we put it into the thermocycler to make sure the top of the plate could get even pressure and heating. This measure reduces the chance of evaporation at the edge of the plate during PCR.

4. In this approach, the initial concentration of one PCR primer is 10 times higher than that of the second PCR primer. At the end of the

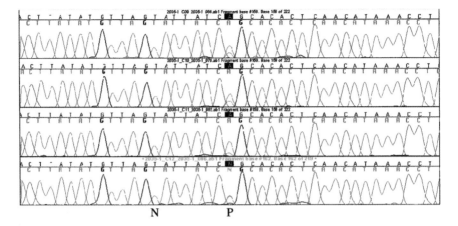

Fig. 2. Allele frequency estimation of SNP in 3 pooled samples. The sequencing traces are arranged as in **Fig. 1**. The alleles are T and A, with the reference individual being a heterozygote. Because there are no A peaks similar in height to the polymorphic A peak in the reference individual (bottom panel), but a T peak very similar in height to the polymorphic T peak, the T peak is chosen as the reference allele (with peak height P). The normalizing peak (with peak height N) is chosen as before. By comparing the ratios of P/N of the pool sample and the reference sample, the allele frequencies are estimated to be 29, 27, and 19% T, respectively, in the African American, Japanese/Chinese, and Caucasian panels. The allele frequencies for A are therefore 71, 73, and 81%, respectively.

PCR, the PCR primer at lower initial concentration is consumed completely, leaving one PCR primer to serve as the sequencing primer. The concentration of the dNTPs used in the PCR step is also reduced so as not to alter the dNTP composition of the sequencing mix drastically. The PCR primer concentration is designed to have sufficient sequencing primer in the sequencing step when only 2.5 µL of the asymmetric PCR product is used in the sequencing reaction. If one needs to sequence for both sense and antisense strands of the PCR product, two PCR reactions—one with higher sense primer concentration and one with higher antisense primer concentration—have to be set up. Our experience shows that this protocol works for PCR products up to 600 bp.

5. To get enough volume of purified samples to run on the capillary DNA sequencer and to ensure full DNA recovery from the Centri-Sep columns, we add 12 µL of water to the sequencing reaction product before it is purified.

6. To use the ABI Prism®3700 DNA analyzer, only 96-well reaction plates provided by Applied Biosystems can be used with the instrument. Therefore, samples have to be transferred to these plates before running the instrument. We seal plates with aluminum thermowell sealers to prevent evaporation while the plate awaits loading onto the capillaries. Centrifuge the plate containing sequencing products to eliminate bubbles before loading it onto the machine. If ABI Prism®377 or 373 DNA Sequencer are used, the purified sequencing reaction product has to be dried in Speed-Vac, and redissolved in 6 µL of loading buffer, heat at 95°C for 2 min, before loading 2.5 µL of the sample onto the gel plate.

7. The peak heights can be measured in two ways. One can print out the sequencing traces and measure the peaks by hand. Alternatively, one can capture the aligned Sequencher chromatograms and export it to a simple graphics program such as "Paint" or "PhotoPlus" where one can measure the heights using an electronic ruler.

8. In choosing the normalizing peak, one looks for the same base type as the reference peak that is not too close to the polymorphic site because the peak heights of the 3 bases before or after the polymorphic site can vary according to the peaks found at the polymorphic site. We also pick the normalizing peak not too far from the polymorphic site because the general peak heights in the vicinity do not vary as much. Finally, if the height of the normalizing peak is very close to the reference peak, any measurement error will not affect the outcome of the estimate too drastically.

9. In a heterozygous reference sequence, the peak height of the reference allele represents a 50% contribution of the allele, one must therefore account for this by multiplying the ratio between the peak heights of the alleles in the pool sequence and the reference sequence by 0.5.

10. Because in most cases, the reference sequence is derived from a homozygote, the second allele cannot be estimated empirically. It is simpler to just subtract the allele frequency of the reference allele from 1 to obtain the estimated allele frequency of the second allele.

Acknowledgments

This work is supported by grants from the National Institutes of Health and the SNP Consortium.

References

1. Taillon-Miller, P. and Kwok, P.-Y. (1999) Efficient approach to unique single nucleotide polymorphism discovery. *Genome Res.* **9**, 499–505.
2. Zakeri, H., Amparo, G., Chen, S.-M., Spurgeon, S., and Kwok, P.-Y. (1998) Peak height pattern in dRhodamine and BigDye terminator sequencing. *Biotechniques* **35**, 406–414.
3. Parker, L. T, Deng, Q., Zakeri, H., Carlson, C., Nickerson, D. A., and Kwok, P.-Y. (1995) Peak height variations in automated sequencing of PCR products using Taq dye-terminator chemistry. *Biotechniques* **19**, 116–121.
4. Parker, L. T., Zakeri, H., Deng, Q., Spurgeon, S., Kwok, P.-Y., and Nickerson, D. A. (1996) AmpliTaq DNA polymerase, FS dye-terminator sequencing: analysis of peak height patterns. *Biotechniques* **21**, 694–699.
5. Kwok, P.-Y., Carlson, C., Yager, T., Ankener, W., and Nickerson, D. A. (1994) Comparative analysis of human DNA variations by fluorescence-based sequencing of PCR products. *Genomics* **23**, 138–144.
6. Rozen S. and Skaletsky, H. (2000) Primer3 on the WWW for general users and for biologist programmers. *Methods Mol. Biol.* **132**, 365–386.
7. Beasley, E. M., Myers, R. M., Cox, D. R., and Lazzeroni, L. C. (1999) Statistical refinement of primer design parameters, in *PCR Applications* (Innis, M. A., Gelfand, D. H., and Sninsky, J. J., ed.), Academic Press, London), pp. 55–72.

7

Computational SNP Discovery in DNA Sequence Data

Gabor T. Marth

1. Introduction

Both the quantity and the distribution of variations in DNA sequence are the product of fundamental biological forces: random genetic drift, demography, population history, recombination, spatial heterogeneity of mutation rates, and various forms of selection. In humans, single base-pair substitution-type sequence variations occur with a frequency of approx 1 in 1.3 kb when two arbitrary sequences are compared *(1)*. This frequency increases with higher sample size *(2)*, i.e., we expect to see, on average, more single nucleotide polymorphisms (SNPs) when a higher number of individual chromosomes are examined *(3,4)*.

SNPs currently in the public repository *(5)* were discovered in DNA sequence data of diverse sources, some already present in sequence databases, but the majority of the data generated specifically for the purpose of SNP discovery. Nearly 100,000 SNPs in transcribed regions were found by analyzing clusters of expressed sequence tags (ESTs) *(6–8)*, or by aligning ESTs to the human reference sequence *(9)*. The three major sources of genomic SNPs were sequences from restricted genome representation libraries *(10)*, ran-

From: *Methods in Molecular Biology, vol. 212:*
Single Nucleotide Polymorphisms: Methods and Protocols
Edited by: P-Y. Kwok © Humana Press Inc., Totowa, NJ

dom shotgun reads aligned to genome sequence *(1)*, and the overlapping sections of the large-insert (mainly bacterial artificial chromosome, or BAC) clones sequenced for the construction of the human reference genome *(11–13)*. Most of these SNPs were detected in pairwise comparisons where one of the two samples was a genomic clone sequence. Theory predicts *(14)*, and experiments confirm, that shallow sampling results in an overrepresentation of common variations: these common SNPs tend to be ancient variations, often present in all or most human populations *(15)* and expected to be valuable for detecting statistical association *(16)*. For the same reason, many rare polymorphisms with rare phenotypic effects are likely to be absent from this set. The current collection of SNPs forms a dense, genome-wide polymorphism map *(1)* intended as a starting point for regional variation studies. An exhaustive survey of polymorphisms in a given region of interest is likely to require significantly higher sample sizes. Even so, the isolation of rare phenotypic mutations may only be possible by the crosscomparison between large samples of affected patients and those of controls.

Computational SNP discovery, in a general sense, refers to the process of compiling and organizing DNA sequences that represent orthologous regions in samples of multiple individuals, followed by the identification of polymorphic sequence locations. The first step typically involves a similarity search with the Basic Local Alignment Search Tool (BLAST) *(17)* to compile groups of sequences that originate from the region under examination. This is followed by the construction of a base-wise multiple alignment to determine the precise, base-to-base correspondence of residues present in each of the samples in a group. Finally, each position of the multiple alignment is scanned for nucleotide mismatches.

Some of the most serious difficulties of sequence organization stems from the repetitive nature of the DNA observed in many organisms. It is well known that nearly half of the human genome is made up of high copy-number repetitive elements *(18,19)*. In addition, many intra- and interchromosomal duplication exist, a large number of them yet uncharacterized. Similar to members of multi-gene families, these duplicated (paralogous) genomic regions may exhibit extremely high levels of sequence similarity *(18)*, sometimes over 99.5%, and can extend over hundreds of kilobases. Failure to distinguish between sequences from different copies of duplicated regions results in false SNP predictions that represent paralogous sequence differences rather than true polymorphisms.

The construction of correct base-wise multiple alignments is a difficult problem because of its computational complexity. Sequences under consideration are generally of different length rendering global sequence alignment algorithms such as CLUSTALW *(20)* rarely applicable. Expressed sequences (ESTs or more or less complete gene sequences) require local alignment techniques that are unperturbed by exon-intron punctuation and alternatively spliced sequence variants.

Once a multiple alignment is constructed, nucleotide differences among individual sequences can be analyzed. Owing to the presence of sequencing errors, not every nucleotide position with mismatches automatically implies a polymorphic site. Although it is impossible to decide which is the case with certainty, the success of SNP detection ultimately depends on how well one is able to discriminate true polymorphisms from likely sequencing errors. This is usually accomplished by statistical considerations that take advantage of measures of sequence accuracy *(21,22)* accompanying the analyzed sequences. The result, ideally, is a set of candidate SNPs, each with an associated SNP score that indicates the confidence of the prediction. Accurate confidence values can be extremely useful for the experimentalist in selecting which SNPs to use in a study or for further characterization, and enables one to use the highest number of candidates within the bounds of an acceptable false positive rate.

2. Materials

Sequences used in SNP analysis come from diverse sources. From the viewpoint of sequence accuracy, they can be categorized as either single-pass sequence reads or consensus sequences that result from multipass, redundant sequencing of the same underlying DNA.

The overall sequencing error rate of single-pass sequences is in the 1%-range *(21–23)*, an order of magnitude higher than the average polymorphism rate (roughly 0.1%). The error rate is typically much higher at the beginning and the end of a read *(21,22)*. Clusters of sequencing errors are also common; the location of these is highly dependent on specific base combinations, as well as the sequencing chemistry used. For detecting sequence variations, even marginally accurate data can be useful as long as regions of low accuracy nucleotides can be avoided. The most widely used base-calling program, PHRED *(21,22)* associates a base quality value to each called nucleotide. This base quality value, Q, is related to the likelihood that the nucleotide in question was determined erroneously: $Q = -10$ $log_{10}(P_{error})$. Although different sequencing chemistries pose different challenges to base calling, tests involving large data sets have demonstrated that the quality value produced by PHRED is a very good approximation of actual base-calling error rates *(21,22)*. Using base quality values, mismatches between low-quality nucleotides can be discarded as likely sequencing errors. Because consensus sequences are the product of multiple sequence reads, they are generally of higher accuracy. Exceptions to this rule are regions where the underlying read coverage is low, and/or regions where all underlying reads are of very low quality. Recognizing this problem, sequence assemblers (computer programs that create consensus sequences) also provide base quality values for the consensus sequence by combining quality scores of the underlying reads *(24,25)*. The following subsections describe the most commonly used sequence sources used in SNP discovery.

2.1. STS Sequences

Sequence-tagged site (STS) sequences, amplified and sequenced in multiple individuals, were used in the first large-scale efforts to catalog variations at the genome scale *(26)*. One of the main advantages of this strategy was that PCR primers, optimized during STS development, were readily available for use. If starting material for the amplification is genomic DNA, these sequences represent the superposition of both copies of a chromosome within an individual. As a result, the sequence may contain nucleotide ambiguities that correspond to heterozygous positions in the individual. Base-calling algorithms trained for homozygous reads will assign a low base quality value to whichever nucleotide is called, rendering base quality value-based SNP detection algorithms ineffective for these reads. Specialized algorithms *(31)* have been designed to deal with heterozygote detection, as discussed next.

2.2. EST Sequences

Expressed Sequence Tag (EST) Reads represent the richest source of SNPs in transcribed regions *(6–8,27,28)* to date. The majority of ESTs are single-pass reads, often from tissue-specific cDNA libraries *(29,30)*. Because a single EST read may contain several exons, special care must be taken when these reads are aligned to genomic sequences. An additional difficulty is the alignment of ESTs representing alternative splice-variants of a single gene.

2.3. Small Insert Clone Sequences

2.3.1. Sequences from Reduced Representation Libraries

Size-Selected Restriction Fragments recognized by specific restriction enzymes are quasirandomly distributed in genomic DNA. The average distance between neighboring restriction sites (restriction fragment length) is a function of the length of the recognition sequence. A reduced, quasirandom representation of the genome can be achieved by first constructing a library of cloned restriction fragments, followed by size-selection to exclude fragments outside a desired length range. The number of different fragments (complexity) present in the library can be precalculated for any given length range. Inversely, library complexity can be controlled by appropriate selection of the upper and lower size limits *(10)*.

2.3.2. Sequences from Random Genomic Shotgun Libraries

Random Genomic Subclone Reads are sequenced from DNA libraries with a quasirandom, short-insert subclone representation of the entire genome (whole-genome shotgun libraries). Because these reads deliver a random sampling of the whole genome, they are well-suited for genome-wide SNP discovery *(1,12)*.

2.4. Large-Insert Genomic Clone Consensus Sequences

Recent large-scale, genome-wide SNP discovery projects *(1,11–13,32)* take advantage of the public human reference sequence built as a tiling path through partially overlapping, large-insert genomic clones *(18,23)*. The sequence of these clones was determined with a local shotgun strategy. By cloning random fragments into a suitable sequencing vector, a subclone library is created for each clone. This library is then extensively sequenced until reaching a desired, three- to tenfold, quasirandom read coverage. The DNA sequence of the large-insert clone is reconstructed by assembling the shotgun reads with computer programs *(24)*. At this stage, there are still several gaps in the sequence, although overall accuracy is high (approx 99.9%). Gap closure and clean up of regions of low-quality sequence requires considerable manual effort *(23)* known as "finishing." Finished or "base-perfect" sequence is assumed at least 99.99% accurate *(18)*.

2.5. Assembled Whole-Genome Shotgun Read Consensus Sequences

Similar in nature to genomic clone sequences, these consensus sequences are the result of assembling a large number of genome-wide shotgun reads, possibly from libraries representing multiple individuals. Over two million human SNP candidates were discovered in the private sector by the analysis of multi-individual reads that provided the raw material for the construction of a human genome reference sequence produced by the whole-genome sequence assembly method *(19)*.

3. Methods

3.1. Published Methods of SNP Discovery

Methods of SNP mining have gone through a rapid evolution during the past few years. The first approaches relied on visual comparison of sequence traces from multiple individuals *(33)*. Although manual comparison of a small number of sequence traces is feasible, standard accuracy criteria are hard to establish, and this method does not scale well for multiple sequence traces and many polymorphic locations. The efficiency of visual inspection is increased when it is performed in the context of a multiple sequence alignment *(27,34,35)*, aided by computer programs that are capable of displaying the alignments and provide tools for simultaneous viewing of sequence traces at a given locus of the multiple alignment *(36)*. Computer-aided prefiltering followed by manual examination of sequence traces *(11,32)* was used in the analysis of overlapping regions of genomic clone sequences to detect candidate SNPs as sequence differences between reads representing the two overlapping clones. These early methods were instrumental in demonstrating the value of extant sequences, sequenced as part of the Human Genome Project, for the discovery of DNA sequence variations. Although visual inspection remains an integral part of software testing and tuning, demands for fast and reliable SNP detection in large data sets have necessitated the development of automated, computational methods of SNP discovery.

The first generation of these methods was designed to enable mining the public EST database *(37)*, and relied, in part, on tools previously developed to aid the automation of DNA sequencing *(23)*. SNP detection was performed by software implementing heuristic considerations. Picoult-Newberg et al. *(27)* used the genome fragment assembler PHRAP to cluster and multiply align ESTs from 19 cDNA libraries. The use of the genome assembler implied that alternatively spliced ESTs were not necessarily included in a single cluster. There was no attempt to distinguish between closely related members of gene families (paralogs). SNP detection was carried out through the successive application of several filters to discard SNP candidates in low-quality regions, followed by manual review. Mainly as the result of conservative heuristics, this method only found a small fraction, 850 SNP candidates in several hundreds of thousands of sequences analyzed. Buetow et al. *(6)* used UNIGENE *(38)*, a collection of precomputed EST clusters as a starting point. ESTs within each cluster were multiply aligned with PHRAP *(24)*. Identification of paralogous subgroups within clusters was done by constructing phylogenetic trees of all cluster members and analyzing the resulting tree topology. Again, SNP candidates were identified by heuristic methods to distinguish between true sequence differences and sequencing errors. This method yielded over 3,000 high-confidence candidates in 8,000 UNIGENE clusters that contained at least 10 sequence members. Unfortunately, the great majority of clusters contained significantly fewer sequences that could not be effectively analyzed with these methods.

The development of a second generation of tools was prompted by the needs of genome-scale projects of SNP discovery. The large amount of data generated by The SNP Consortium (TSC) *(1)* has spurred the development of several SNP discovery tools. In the initial phase, the TSC employed a molecular strategy called restricted genome representation (RRS), which involves the sequencing of size-selected restriction fragment libraries from multiple individuals *(10)*. For example, the full digestion by a given restriction enzyme may produce 20,000 genomic fragments in the 450–550-bp length range. After digestion of the genomic DNA of each of the 24 individuals, followed by size-selection, the restriction fragment libraries are pooled. When a collection of such random fragments is sequenced to appreciable redundancy (say, 60,000–80,000 reads), the sequence of many of the fragments will be available from more than one individual. These redundant sequences are a suitable substrate for SNP analysis. The analysis of data of this type is similar to that of EST sequences. First, one must cluster the sequence reads to delineate groups of identical fragments. To avoid grouping sequences based on similarity between known human repeats they

bly program such as PHRAP) it is possible to objectively and simultaneously evaluate all available data present in the alignment, without regard to sequence source or restrictions on data quality. For each site of the alignment, the algorithm outputs the probability that the site is polymorphic. These probability values were shown to accurately estimate the validation rate of candidate SNPs in various mining applications *(1,9,15)*. This is desirable because realistic estimates for the true positive rate allow one to use the highest number of SNP candidates within an acceptable false positive rate. The POLYBAYES software is compatible with the PHRED/PHRAP/CONSED file structure, is capable of analyzing multiple alignments created with PHRAP, and the output, including markup information such as paralog tags and candidate SNP sites, is directly viewable within CONSED (**Figs. 2** and **3**). An alternative statistical formulation *(8)* developed to analyze EST clusters produces a log-odds (LOD) score to rank SNP candidates based on sequence accuracy, the quality of the alignment, prior polymorphism rate, and by evaluating adherence to the rules of Mendelian segregation of alleles within individual cDNA libraries.

There are two additional cases of practical importance that the algorithms described earlier were not designed to work with directly. In many situations, the DNA template that is available for analysis is double stranded, genomic DNA of an individual, or sometimes a pool of multiple individuals. The first is the case when a known region is assayed from the genomic DNA of multiple individuals *(34,35)*, giving rise to sequence traces that contain heterozygous nucleotides. An example of a multi-individual DNA pool is one constructed to obtain population-specific estimates of allele frequency of known polymorphisms *(42)*. PCR products obtained from such starting material represent more than a single, unique strand of DNA. When these products are sequenced, polymorphic locations between different strands of DNA appear as base ambiguities in the sequence trace (**Fig. 4**). The automation of heterozygote detection motivated the development of POLYPHRED *(31)*, a computer program *(43)* that examines numerical characteristics of sequence traces such as drop in peak-height, ratio of a second peak under the

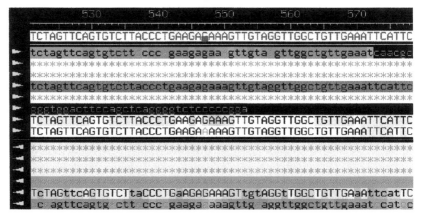

Fig. 3. Candidate SNP site. The SNP (alleles A/G) is evident within members of one of the two alternatively spliced forms of ESTs aligned to the genomic anchor sequence at this location. The tag above, generated automatically by the detection software POLYBAYES, shows the most likely allele combination at the site, together with the probability of that variation.

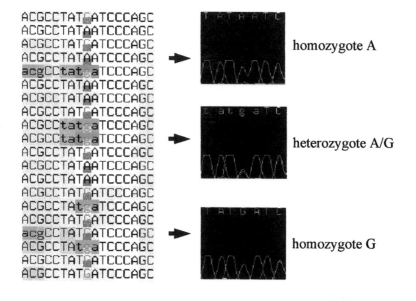

primary peak, and overall sequence quality in the neighborhood of the analyzed nucleotide position. POLYHRED integrates seamlessly with the University of Washington PHRED/PHRAP/CONSED genome analysis software package. Although both POLYPHRED, and other specialized, heuristic approaches has been tested for allele frequency estimation in pooled sequencing, reliable computer algorithms of frequency estimation are not yet available.

Another topic of practical importance is the detection of short insertions and deletions (INDELs). Polymorphisms of this type are also commonly referred to as DIPs (deletion-insertion polymorphisms). The main difficulty of detecting DIPs is the fact that current, base-wise measures of sequence accuracy provide no direct estimates of insertion or deletion type sequencing errors. The base quality value, accompanying a given nucleotide, expresses the likelihood that the nucleotide was called in error, but it is not possible to separate the likelihood of substitution-type sequencing error from the likelihood that a nonexistent nucleotide was artifactually inserted by the base caller. Similarly, there is no direct measure of the likelihood that between two called, neighboring nucleotides there are additional bases in the sequencing template that were erroneously omitted and therefore represent deletion-type errors. In the absence of sequencing error estimates, it is difficult to formulate rigorous models of insertion-deletion type polymorphisms. A heuristic approach employed by POLYBAYES for DIP detection is based on the assumptions that a higher base quality value corresponds to a decreased chance that the called nucleotide is, in fact, an artifactual insertion, and that the likelihood of deleted nucleotides

Fig. 4. Heterozygote detection with the POLYPHRED program. Multiple alignment with the site of an SNP marked up with POLYPHRED (left). Sequence traces of a homozygous A/A, a heterozygous A/G, and a homozygous G/G individual (right).

between two high-quality called bases is low. Taking into account the base quality value of the nucleotides neighboring a candidate deletion, as well as the base quality values of the corresponding candidate insertion in another aligned sequence, a heuristic DIP likelihood is calculated. This likelihood was used to detect DIPs in overlapping regions of large-insert clones of the Human Genome Assembly. Validation rate for DIPs that were at least two base pairs long was about 70%; the validation rate for single base-pair insertions-deletions was significantly lower, especially for base-number differences in mono-nucleotide runs.

3.2. Computational Aspects of SNP Discovery

The majority of software packages for automated SNP discovery were developed to run under the UNIX operating system. Part of the reason for this is the availability of powerful and flexible programming tools that UNIX provides for the software developer. In addition, many of the SNP discovery tools available today were written in a way that enables their integration into existing genome analysis packages such as the PHRED/PHRAP/CONSED system, developed at the University of Washington under UNIX. Hardware requirements for SNP mining depend greatly on the scope of the task tackled. Searching for SNPs in specific, short (up to 100–150 kb) regions of the genome, in up to a few hundred sequences, is well within the capabilities of a conventional UNIX workstation (or a computer running the user-friendly LINUX operating system that can be installed on a personal computer with relative ease). Genome-wide SNP mining projects typically require server-class machines, and access to several hundred gigabytes of data storage, especially if intermediate steps of the mining procedure are tracked and results are recorded in a database.

Unfortunately, there is no official standard data exchange format for sequence multiple alignments, or SNP markup information. Many of the SNP discovery tools currently in use expect input and produce output in file formats specific to the program. In these cases, data translation between different tools is achieved via custom scripts. The closest to a *de facto* standard is the PHRED/PHRAP/CONSED *(24)* file structure and software architecture developed at the University of Washington that is widely used in sequencing laboratories worldwide. Given that several of the main SNP analysis tools, including POLYPHRED and POLYBAYES, were built to integrate within this structure, it is worthwhile to briefly summarize the University of Washington package standards for representing SNP information.

The main directory of the file architecture contains four subdirectories in which all relevant data is organized. Sequence traces reside in the subdirectory **chromat_dir**. When the base-calling algorithm PHRED interprets a trace, it creates a sequence analysis file in the PHD format, and writes it into the subdirectory **phd_dir**. In addition to header information such as sequence name, read chemistry, and template identifier, the PHD format file contains three important pieces of information for each called base: the called DNA residue, the corresponding base quality value describing the accuracy of the call, and the position of the called nucleotide relative to the sequence trace. The PHD file may also contain permanent additional sequence information or tags attached to sections of the read (such as the region of an annotated repeat, or cloning vector sequence). The pre-requisite of using POLYPHRED is the presence of an additional trace analysis file that contains detailed information about the trace, at the location of the called nucleotide. This file is the POLY format trace analysis file, located in the subdirectory **poly_dir**. Finally, all downstream analysis files are kept in the fourth subdirectory **edit_dir**. Perhaps the most commonly used file in this directory is the ACE format sequence assembly, or multiple alignment file. This file format was designed as an interchange format between the PHRAP sequence assembly program and the CONSED sequence editor. ACE files are versioned and sequence edits performed within CONSED are saved as consecutive versions. The SNP detection program POLYPHRED takes an ace format multiple alignment file, and adds markup information regarding the location of heterozygous trace positions. These tags are visible when the alignment is viewed with CONSED, enabling rapid manual review. POLYBAYES operates in one of two modes. The first mode is the analysis of a pre-existing multiple alignment, supplied in the ACE format. In this case, the anchored multiple alignment step is bypassed, and an ACE format output file is created that contains the results of paralog identification and SNP detection, again, as tags viewable from within CONSED. In the second mode of operation one utilizes the anchored alignment capability of POLYBAYES. In this case, one starts out with FASTA format files representing the DNA sequence and the accompanying base quality values for the genomic anchor sequence, as well as the cluster member sequences (for a description of the FASTA format *see* URL: http://www.ncbi.nlm.nih.gov/BLAST/fasta.html). CROSS_MATCH *(24)*, a pair-wise, dynamic programming alignment algorithm is run between each member sequence and the anchor. The sequences, together with the pair-wise alignmentsare supplied to POLYBAYES. The program multiply aligns the member sequences, performs the paralog filtering and the SNPdetection step, and produces a new ACE format output file for the viewing of the anchored multiple alignment and SNP analysis results.

3.3. SNP Discovery Protocol

Given the diversity of sequence data that can be used to detect polymorphic sites within an organism, it is impossible to prescribe a single protocol that works in every situation. In general, the mining procedure will contain the following steps: data organization, the creation of a base-wise multiple alignment, filtering of paralogous sequences (or cluster refinement), followed by the detection of SNPs in slices of the multiple alignment. In this final section of this chapter, we will give two different examples that typify the usual steps of SNP mining. The majority of mining applications can be successfully completed by customizing and combining these steps.

3.3.1. SNP Discovery in EST Sequences

In the first scenario, in a screen against a cDNA library one pulls out a clone sequence that contains a gene of interest. The cDNA is an already sequenced clone, the corresponding EST is in the public database, dbEST (37) (URL: http://www.ncbi.nlm.nih.gov/dbEST). The goal is to explore single base-pair variations within the gene. The first step towards this goal is to find all SNPs in those transcribed sequences of the gene that are available in public sequence databases. One proceeds as follows:

1. Find the location of the gene in the human genome from which the EST was expressed. Go to the NCBI (National Center for Biotechnology Information) web site (URL: http://www.ncbi.nlm.nih.gov) and follow the Map Viewer link. Use the search facility on this page to find the genomic location of the EST, pre-computed by the NCBI. Perform the search using the accession number of the EST. Make sure that you set the "Display Settings" to include the "GenBank" view. Click on the genome clone accession that overlaps the EST, and download the sequence in FASTA format. This sequence will act as the genomic anchor sequence for the ESTs to be analyzed.

2. Find all other ESTs in dbEST with significant sequence similarity to the original EST sequence. Perform the similarity search from the NCBI (National Center for Biotechnology Information) website (URL: http://www.ncbi.nlm.nih.gov/BLAST). Choose the "Standard nucleotide-nucleotide BLAST" option. Type the accession number of the EST in the "Search" field. Choose "est_human" as the database to search against. Once the search is done, format the output as "Simple text," and parse out the accession list of ESTs from the list of hitting sequences (*see* **Note 1**).

3. Retrieve EST sequence traces. In the near future, EST trace retrieval will be possible from the trace repository (URL: http://www.ncbi.nlm.nih.gov/Traces) that is under construction at the NCBI. Currently, EST sequence traces can be downloaded from the Washington University ftp site: (URL: ftp://genome.wustl.edu/pub/gsc1/est) for ESTs produced there. Searching is done via the local EST names. Download all ESTs for which traces can be found at this site (*see* **Note 2**).

4. Process the sequence traces with the PHRED base-calling program. Invoke PHRED with the command line parameters that produce files necessary for downstream processing in the University of Washington PHRED/PHRAP/CONSED architecture (URL: http://www.phrap.org). Make sure that PHD format sequence files are created in the "phd_dir" subdirectory, by specifying the location of this directory with the "-cd" option. Use the utility program PHD2FASTA (provided with CONSED) to produce a FASTA format file of the DNA sequences ("-os" option) of the ESTs file. Also, produce a FASTA format file for the accompanying base quality values ("-oq" option), and one for the list of base positions that specify the location of each called nucleotide relative to the sequence trace ("-ob" option). The DNA sequence of the ESTs will be used in the next step, as the members of the cluster (group) of expressed sequences to analyze for polymorphic sites.

5. Create a multiple alignment of the EST sequences with the anchored alignment algorithm implemented within POLYBAYES (instructions at the POLYBAYES web site, URL: http://genome.wustl.edu/gsc/polybayes). As the anchor sequence, use the genomic clone sequence from **step 1**. Use the CROSS_MATCH dynamic alignment program to compute the initial pair-wise alignments between each of the ESTs and the genomic anchor sequence (CROSS_MATCH is distributed as part of the PHRAP software package *[24]*). As cluster member sequences, use the ESTs obtained in **steps 2–4**. **Figure 1** shows a section of a sample multiple alignment, viewed with the CONSED *(36)* sequence viewer-editor program. Observe that, in this case, the ESTs are divided into two groups of alternative splice forms.

6. Likely paralogous sequences are identified with the in-built paralog-filtering feature of POLYBAYES. This feature is invoked by the "-filterParalogs" command line option (additional relevant arguments explained in the online documentation available at the POLYBAYES web site). **Figure 2** shows a different section of the multiple alignment produced in the previous step. Observe that there are several high-quality mismatches between the genomic anchor sequence and EST marked with the blue tag. This sequence is considered a sequence paralog, and is automatically tagged by the filtering algorithm. The paralogous sequence is removed from consideration in any further analysis.

7. The multiple alignment is scanned for polymorphic sites. At each site, the slice of the alignment composed of nucleotides contributed

27. Picoult-Newberg, L., Ideker, T. E., Pohl, M. G., Taylor, S. L., Donaldson, M. A., Nickerson, D. A., and Boyce-Jacino, M. (1999) Mining SNPs from EST databases. *Genome Res.* **9**, 167–174.

28. Garg, K., Green, P., and Nickerson, D. A. (1999) Identification of candidate coding region single nucleotide polymorphisms in 165 human genes using assembled expressed sequence tags. *Genome Res.* **9**, 1087–1092.

29. Hillier, L. D., Lennon, G., Becker, M., Bonaldo, M. F., Chiapelli, B., Chissoe, S., et al. (1996) Generation and analysis of 280,000 human expressed sequence tags. *Genome Res.* **6**, 807–828.

30. Adams, M. D., Soares, M. B., Kerlavage, A. R., Fields, C., and Venter, J. C. (1993) Rapid cDNA sequencing (expressed sequence tags) from a directionally cloned human infant brain cDNA library. *Nat. Genet.* **4**, 373–380.

31. Nickerson, D. A., Tobe, V. O., and Taylor, S. L. (1997) PolyPhred: automating the detection and genotyping of single nucleotide substitutions using fluorescence-based resequencing. *Nucleic Acids Res.* **25**, 2745–2751.

32. Dawson, E., Chen, Y., Hunt, S., Smink, L. J., Hunt, A., Rice, K., et al. (2001) A SNP resource for human chromosome 22: extracting dense clusters of SNPs from the genomic sequence. *Genome Res.* **11**, 170–178.

33. Kwok, P.-Y., Carlson, C., Yager, T. D., Ankener, W., and Nickerson, D. A. (1994) Comparative analysis of human DNA variations by fluorescence-based sequencing of PCR products. *Genomics* **23**, 138–144.

34. Nickerson, D. A., Taylor, S. L., Weiss, K. M., Clark, A. G., Hutchinson, R. G., Stengard, J., et al. (1998) DNA sequence diversity in a 9.7-kb region of the human lipoprotein lipase gene. *Nat. Genet.* **19**, 233–240.

35. Nickerson, D. A., Taylor, S. L., Fullerton, S. M., Weiss, K. M., Clark, A. G., Stengard, J. H., et al. (2000) Sequence diversity and large-scale typing of SNPs in the human apolipoprotein E gene. *Genome Res.* **10**, 1532–1545.

36. Gordon, D., Abajian, C., and Green, P. (1998) Consed: a graphical tool for sequence finishing. *Genome Res.* **8**, 195–202.

37. Boguski, M. S., Lowe, T. M., and Tolstoshev, C. M. (1993) dbEST: database for "expressed sequence tags". *Nat. Genet.* **4**, 332–333.

38. Wheeler, D. L., Church, D. M., Lash, A. E., Leipe, D. D., Madden, T. L., Pontius, J. U., et al. (2001) Database resources of the National Center for Biotechnology Information. *Nucleic Acids Res.* **29**, 11–16.

39. Smit, A. F. A. G., P., http://ftp.genome.washington.edu/RM/ RepeatMasker.html

40. Ning, Z., Cox, A. J., and Mullikin, J. C. (2001) SSAHA: A fast search method for large DNA databases. *Genome Res.* **11**, 1725–1729.

41. Collins, F. S., Patrinos, A., Jordan, E., Chakravarti, A., Gesteland, R., and Walters, L. (1998) New goals for the U.S. Human Genome Project: 1998–2003. *Science* **282**, 682–689.

42. Kwok, P.-Y. (2000) Approaches to allele frequency determination. *Pharmacogenomics* **1**, 231–235.

43. Nickerson, D. A., http://droog.mbt.washington.edu/PolyPhred.html

8

Genotyping SNPs With Molecular Beacons

Salvatore A. E. Marras, Fred Russell Kramer,
and Sanjay Tyagi

1. Introduction

Single-nucleotide substitutions represent the largest source of diversity in the human genome. Some of these variations have been directly linked to human disease, though the vast majority are neutral. Even neutral variations are important because they provide guideposts in the preparation of detailed maps of the human genome, serving as essential elements in linkage analyses that identify genes responsible for complex disorders *(1)*. Although sequencing is adequate for the initial discovery of single-nucleotide variations, simpler, faster, and more automated genotyping methods are needed for routine clinical diagnostics and population studies. High-throughput methods are essential for understanding the distribution of genetic variations in populations, as well as for identifying the genes responsible for genetic disorders. Current alternatives to sequence analysis either miss some single-nucleotide substitutions or are too complex to enable high-throughput assays *(2)*.

We have developed a simple method for typing single nucleotide polymorphisms in which alleles are identified by fluorescent colors generated in sealed amplification tubes. In this technique, amplifi-

From: *Methods in Molecular Biology, vol. 212:*
Single Nucleotide Polymorphisms: Methods and Protocols
Edited by: P-Y. Kwok © Humana Press Inc., Totowa, NJ

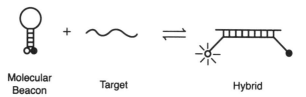

Fig. 1. Principle of operation of molecular beacons. When the probe sequence in the loop of a molecular beacon binds to a target sequence a conformational reorganization occurs that restores the fluorescence of a quenched fluorophore.

cation is carried out in the presence of two different molecular beacons and each allele is identified by the development of fluorescence of a unique color. Molecular beacons are single-stranded oligonucleotide probes that become fluorescent when they bind to perfectly complementary nucleic acids *(3)*. Because they are non-fluorescent when they are not bound to their target, they can be used in hybridization reactions without having to separate the probe-target hybrids from the nonhybridized probes. Molecular beacons possess a stem-and-loop structure. A fluorophore is covalently linked to one end of the molecule and a quencher is covalently linked to the other end. When not bound to target, the hairpin stem keeps the fluorophore so close to the quencher that fluorescence does not occur. The energy absorbed by the fluorophore is transferred to the quencher and released as heat. However, when the probe sequence in the loop anneals to its target sequence, the rigidity of the probe-target hybrid forces the hairpin stem to unwind, separating the fluorophore from the quencher, and restoring fluorescence (*see* **Fig. 1**). Because molecular beacons can possess a wide variety of differently colored fluorophores, multiple targets can be distinguished in the same solution, using several different molecular beacons, each designed to detect a different target, and each labeled with a different fluorophore *(4–6)*.

The ability of molecular beacons to report the presence of their targets without having to isolate the probe-target hybrids makes them useful in a wide array of applications. They can be used to monitor the progress of polymerase chain reactions (PCR) *(4,7)* and nucleic acid sequence-based amplification reactions in sealed tubes *(8,9)*. They can be used to construct self-reporting oligonucleotide arrays *(10)* and to detect specific mRNA sequences in living cells *(11,12)*.

Molecular beacons are uniquely suited for the detection of single-nucleotide variations because they recognize their targets with significantly higher specificity than conventional oligonucleotide probes *(4,7,13)*. Their high specificity is a consequence of their hairpin structure *(13)*. When a molecular beacon binds to its target sequence, the formation of the probe-target hybrid occurs at the expense of the stem hybrid. Molecular beacons can be designed so that over a wide range of temperatures only perfectly complementary probe-target hybrids are sufficiently stable to force open the stem hybrid. Mismatched probe-target hybrids do not form, except at substantially lower temperatures *(7,13)*. Therefore, a relatively wide range of temperatures exist in which perfectly complementary probe-target hybrids elicit a fluorogenic response, while mismatched molecular beacons remain dark. Consequently, assays using molecular beacons robustly discriminate targets that differ from one another by as little as a single nucleotide substitution.

A number of laboratories have confirmed the utility of using molecular beacons for the detection of single nucleotide polymorphisms (SNP). Molecular beacons have been used to genotype Factor V Leiden mutations *(14)*, hereditary haemochromatosis gene mutations *(14)*, cystic fibrosis and Tay-Sachs disease gene mutations *(15)*, methylenetetrahydrofolate gene mutations *(16)*, human chemokine receptor mutations *(17,18)*, drug-resistance mutations in malarian parasites *(19)*, and drug-resistance mutations in *Mycobacterium tuberculosis (20)*. Usually the mutations are detected in real-time during amplification, but they can also be detected after amplification *(21)*. In side-by-side comparisons, the specificity of molecular beacons has proven superior to probes that rely on the 5'-endonucleolytic cleavage activity of DNA polymerase *(22)*. This high specificity allows detection of a small proportion of mutant DNA in the presence of an abundant wild-type DNA *(23)*.

2. Materials

2.1. Molecular Beacons

Molecular beacons can be synthesized by the researcher (*see* Methods section below) or by a commercial supplier. A number of oligonucleotide synthesis companies specialize in the synthesis, purification, and characterization of molecular beacons. These companies include, Biolegio BV (The Netherlands), Biosearch Technologies (CA), BioSource International (CA), Eurogentec (Belgium), Gene Link (NY), Genset Oligos (France), Integrated DNA Technologies (IA), Isogen Bioscience (The Netherlands), Midland Certified Reagents (TX), MWG-Biotech (Germany), Operon Technologies (CA), Oswel (Great Britain), Sigma-Genosys (TX), Synthegen (TX), Synthetic Genetics (CA), TIB MOLBIOL (Germany), and TriLink BioTechnologies (CA).

2.2. Sequences of Molecular Beacons and Primers

The experimental example that is used to illustrate the principals of the method utilize a pair of molecular beacons that were designed to type a C→T transition at position 627 of the human chemokine receptor 2 gene *(18)*. The sequence of the wild-type-specific molecular beacon is 5' FAM-<u>CGC ACC</u> TCT GGT CTG AAG GTT TAT T <u>GGT GCG</u>-DABCYL 3' and the sequence of the mutant-specific molecular beacon is 5' TET-<u>CGC ACC</u> TCT GGT CTG **AAA** GTT TAT T <u>GGT GCG</u>-DABCYL 3', where the underlined nucleotides identify the arm sequences and the bold letters identify the site of the polymorphism. The sequences of the primers that were used are: 5' AGA TGA ATG TAA ATG TTC TTC TAG 3' and 5' CTT TTT AAG TTG AGC TTA AAA TAA GC 3'.

2.3. Choice of Fuorophores for Different Real-Time Instruments

A number of instruments that can monitor the progress of a PCR by fluorescence have recently become available. Because the typing of each SNP is performed using two molecular beacons, where one molecular beacon is specific for the wild-type sequence and the other is specific for the mutant sequence, it is important to carefully select two fluorophores that the available instrument is able to distinguish reliably. The Prism 7700 Sequence Detector spectrofluorometric thermal cycler (Applied Biosystem) uses a blue argon-ion laser (488 nm) as its light source, records the emission spectrum in the range of 500–600 nm, and then computes the fraction of each fluorophore that is present in each tube using a deconvolution algorithm. This instrument is able to excite and discriminate FAM and TET very well. The Smart Cycler (Cepheid), the iCycler (Bio-Rad), and the Mx4000 (Stratagene) utilize either multicolored or white light sources in combination with specific filter sets that allow detection of up to four different fluorophores. These instruments rely on fluorophores that have widely separated spectra with minimum cross-talk, rather than relying on the deconvolution of the emission spectra of closely related fluorophores. With these instruments, either FAM or TET, which emit in the green range, can be used for one molecular beacon and either tetramethylrhodamine, Alexa 546, Cy3, ROX, Texas red, or Cy5, which emit in the red range, can be used for the other molecular beacon. The Light Cycler (Roche Diagnostics) utilizes a blue light-emitting diode and detects fluorescence in either the green or the red range. This instrument can therefore be used with a fluorescein labeled molecular beacon that emits in the green range and a Cy5 or "Light Cycler" red-labeled wavelength-shifting molecular beacon *(5)* that emits in the red range (but is excited by the blue light-emitting diode).

3. Methods

3.1. Synthesis of Molecular Beacons

Conventionally, molecular beacons were synthesized by the manual coupling of fluorophore and quencher moieties to oligonucleotides containing amino and sulfhydryl functionalities at each of their ends. Since the development of phosphoramidites linked to fluorophores and quenchers and controlled-pore glass (CPG) columns, containing covalently linked fluorophores and quenchers, molecular beacons can be made entirely by automated synthesis *(24)*. However, phosphoramidites are still not available for most fluorophores. Mixed synthesis can be performed in order to utilize these fluorophores. In the mixed synthesis mode, the quencher moiety dabcyl is introduced during DNA synthesis and the fluorophore is coupled manually to an amino or a sulfhydryl group. Usually iodoacetamide or maleimide derivatives of fluorophores are used for coupling with sulfhydryl groups and succinimidyl ester derivatives of fluorophores are used for coupling with amino groups. Although the nonfluorescent dye dabcyl has the ability to quench all fluorophores in molecular beacons, a number of other nonfluorescent quenchers, such as the "Black Hole" quenchers and QSY-7, have recently been introduced and found to be effective. DNA synthesis reagents for the incorporation of dabcyl can be obtained from Glen Research and Biosearch Technologies, Black Hole quenchers are available from Biosearch Technologies, and QSY-7 is available from Molecular Probes.

1. Start the DNA synthesis on a 3'-dabcyl CPG column of appropriate size (*see* **Note 1**). After incorporating all the nucleotides, introduce either a 5'-fluorophore, a 5'-sulfhydryl group, or a 5'-amino group. The 5' modifiers should remain protected during synthesis to allow purification of the oligonucleotide in the presence of the protective trityl moiety. Perform the post synthetic steps according to the guidelines specific to each modifier. Dissolve the oligonucleotide in 600 μL buffer A (0.1 *M* triethylamonium acetate, pH 6.5) (*see* **Note 2**).

2. Purify the tritylated dabcyl containing oligonucleotide by high-performance liquid chromatography (HPLC) on a C-18 reverse-phase column, utilizing a linear elution gradient of 20–70% buffer B (0.1 *M* triethylamonium acetate in 75% acetonitrile, pH 6.5) in buffer A that forms over 25 min at a flow rate of 1 mL/min. Monitor the absorption of the elution stream at 260 nm (for DNA) and at 491 nm (for dabcyl). Collect the major peak that has a much higher absorption at 260 nm than at 491 nm and that appears near the end of the elution (due to the presence of its hydrophobic trityl moiety). Consult typical chromatograms that are available on the internet at http://www.molecular-beacons.org.

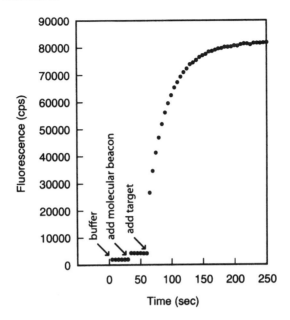

Fig. 2. Functional characterization of a molecular beacon preparation by addition of a complementary oligonucleotide target. There was a 35-fold increase in the fluorescence of this preparation of molecular beacons upon the addition of the target.

molecular beacon, but that does not possess sequences complementary to the arms of the molecular beacon), and monitor the rise in fluorescence until it reaches a stable level (F_{open}).

4. Calculate the signal-to-background ratio as $(F_{open}-F_{buffer})/(F_{closed}-F_{buffer})$.

3.4. Determination of Thermal Denaturation Profiles

In order to determine the window of discrimination, which is the range of temperatures in which perfectly complementary probe-target hybrids can form and in which mismatched probe-target hybrids cannot form, the fluorescence of solutions of molecular beacons in the presence of each kind of target is measured as a function of temperature (*see* **Fig. 3**). This experiment is also useful for confirming the theoretical predictions of different melting transitions.

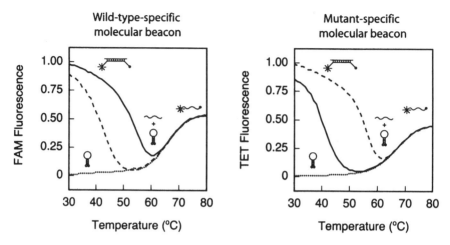

Fig. 3. Thermal denaturation profiles of molecular beacons in the presence of either wild-type target (continuous lines), mutant target (dashed lines), or no target (dotted lines). The state of the molecular beacons is indicated by a diagram over the thermal denaturation profiles. Mismatched hybrids denature 10–12°C below the melting temperature of perfectly matched hybrids. Optimum discrimination is achieved at the center of this temperature range. Therefore, in this example 50°C was chosen to be the assay temperature (the annealing temperature of the PCR).

1. For each molecular beacon, prepare three tubes containing 200 nM molecular beacon, 4 mM MgCl$_2$, 50 mM KCl, and 10 mM Tris-HCl, pH 8.0, in a 50-μL volume.
2. Add a twofold molar excess of an oligonucleotide that is perfectly complementary to the molecular beacon probe sequence (but not to the arm sequences arms) to one of the tubes, add a twofold excess of an oligonucleotide that contains the mismatched target sequence to the other the tube, and add only buffer to the third tube.
3. Determine the fluorescence of each solution as a function of temperature, using a Prism 7700 Sequence Detector spectrofluorometric thermal cycler. Decrease the temperature of the solutions from 80°C to 30°C in 1°C steps, with each step lasting 1 min, while monitoring fluorescence during each step.

3.5. Real-Time PCR

1. Prepare 50-μL reactions that contain 100 nM wild-type-specific molecular beacon, 100 nM mutant-specific molecular beacon, 500 nM

References

1. Sachidanandam, R., Weissman, D., Schmidt, S. C., Kakol, J. M., Stein, L. D., Marth, G., et al. (2001) A map of human genome sequence variation containing 1.42 million single nucleotide polymorphisms. *Nature* **409**, 928–933.
2. Cotton, R. (ed.) (1997) *Mutation Detection.* Oxford University Press, Oxford, UK.
3. Tyagi, S. and Kramer, F. R. (1996) Molecular beacons: probes that fluoresce upon hybridization. *Nat. Biotechnol.* **14**, 303–308.
4. Tyagi, S., Bratu, D. P., and Kramer, F. R. (1998) Multicolor molecular beacons for allele discrimination. *Nat. Biotechnol.* **16**, 49–53.
5. Tyagi, S., Marras, S. A., and Kramer, F. R. (2000) Wavelength-shifting molecular beacons. *Nat. Biotechnol.* **18**, 1191–1196.
6. Vet, J. A., Majithia, A. R., Marras, S. A., Tyagi, S., Dube, S., Poiesz, B. J., and Kramer, F. R. (1999) Multiplex detection of four pathogenic retroviruses using molecular beacons. *Proc. Natl. Acad. Sci. USA* **96**, 6394–6399.
7. Marras, S. A., Kramer, F. R., and Tyagi, S. (1999) Multiplex detection of single-nucleotide variations using molecular beacons. *Genet. Anal.* **14**, 151–156.
8. Leone, G., van Schijndel, H., van Gemen, B., Kramer, F. R., and Schoen, C. D. (1998) Molecular beacon probes combined with amplification by NASBA enable homogeneous, real-time detection of RNA. *Nucleic Acids Res.* **26**, 2150–2155.
9. de Baar, M. P., Timmermans, E. C., Bakker, M., de Rooij, E., van Gemen, B., and Goudsmit, J. (2001) One-tube real-time isothermal amplification assay to identify and distinguish human immunodeficiency virus type 1 subtypes A, B, and C and circulating recombinant forms AE and AG. *J. Clin. Microbiol.* **39**, 1895–1902.
10. Steemers, F. J., Ferguson, J. A., and Walt, D. R. (2000) Screening unlabeled DNA targets with randomly ordered fiber-optic gene arrays. *Nat. Biotechnol.* **18**, 91–94.
11. Matsuo, T. (1998) In situ visualization of messenger RNA for basic fibroblast growth factor in living cells. *Biochim. Biophys. Acta.* **1379**, 178–184.
12. Sokol, D. L., Zhang, X., Lu, P., and Gewirtz, A. M. (1998) Real time detection of DNA.RNA hybridization in living cells. *Proc. Natl. Acad. Sci. USA* **95**, 11,538–11,543.

13. Bonnet, G., Tyagi, S., Libchaber, A., and Kramer, F. R. (1999) Thermodynamic basis of the enhanced specificity of structured DNA probes. *Proc. Natl. Acad. Sci. USA* **96,** 6171–6176.

14. Hu X., Belachew B., Chen L., Huang H., and Zhang J. (2000) Fluoresence-based single-tube assays to rapidly detect human gene mutations. *Stratagies* **13,** 71–73.

15. Fung, C., Tyagi, S., Harris, L., Weisberg, S., Pinter, A., and Kramer, F. R. (2002) Genetic screening using molecular beacons. *Clin. Chem.* **47,** in preparation.

16. Giesendorf, B. A., Vet, J. A., Tyagi, S., Mensink, E. J., Trijbels, F. J., and Blom, H. J. (1998) Molecular beacons: a new approach for semiautomated mutation analysis. *Clin. Chem.* **44,** 482–486.

17. Kostrikis, L. G., Huang, Y., Moore, J. P., Wolinsky, S. M., Zhang, L., Guo, Y., et al. (1998) A chemokine receptor CCR2 allele delays HIV-1 disease progression and is associated with a CCR5 promoter mutation. *Nat. Med.* **4,** 350–353.

18. Gonzalez, E., Bamshad, M., Sato, N., Mummidi, S., Dhanda, R., Catano, G., et al. (1999) Race-specific HIV-1 disease-modifying effects associated with CCR5 haplotypes. *Proc. Natl. Acad. Sci. USA* **96,** 12,004–12,009.

19. Durand, R., Eslahpazire, J., Jafari, S., Delabre, J. F., Marmorat-Khuong, A., di Piazza, J. P., and Le Bras, J. (2000) Use of molecular beacons to detect an antifolate resistance-associated mutation in *Plasmodium falciparum. Antimicrob. Agents Chemother.* **44,** 3461–3464.

20. Piatek, A. S., Tyagi, S., Pol, A. C., Telenti, A., Miller, L. P., Kramer, F. R., and Alland, D. (1998) Molecular beacon sequence analysis for detecting drug resistance in *Mycobacterium tuberculosis. Nat. Biotechnol.* **16,** 359–363.

21. Vogelstein, B. and Kinzler, K. W. (1999) Digital PCR. *Proc. Natl. Acad. Sci. USA* **96,** 9236–9241.

22. Täpp, I., Malmberg, L., Rennel, E., Wik, M., and Syvänen, A. C. (2000) Homogeneous scoring of single-nucleotide polymorphisms: comparison of the 5'-nuclease TaqMan assay and molecular beacon probes. *Biotechniques* **28,** 732–738.

23. Szuhai, K., Ouweland, J., Dirks, R., Lemaitre, M., Truffert, J., Janssen, G., et al. (2001) Simultaneous A8344G heteroplasmy and mitochondrial DNA copy number quantification in myoclonus epilepsy and ragged-red fibers (MERRF) syndrome by a multiplex molecular beacon based real-time fluorescence PCR. *Nucleic Acids Res.* **29,** E13.

24. Mullah, B. and Livak, K. (1999) Efficient automated synthesis of molecular beacons. *Nucleosides Nucleotides* **18,** 1311–1312.

9

SNP Genotyping by the 5'-Nuclease Reaction

Kenneth J. Livak

1. Introduction

The chief attribute of the fluorogenic 5' nuclease assay is that it is completely homogeneous. After mixing the sample and reaction components, the assay is run in a closed tube format with no post-polymerase chain reaction (PCR) processing steps. Results are obtained by simply measuring the fluorescence of the completed reactions. By eliminating post-PCR processing, allelic discrimination with fluorogenic probes reduces the time of analysis, eliminates the labor and supply costs of post-PCR steps, reduces the risk of crossover contamination, and minimizes sources of error. The assay has the sensitivity of PCR so that a minimum amount of genomic DNA is required. The use of endpoint fluorescence measurements maximizes throughput. Using a single ABI PRISM® 7900HT Sequence Detection System and 24-h operation, it is possible to generate up to 250,000 SNP results per day. TaqMan® MGB probes and the entire operating system outlined below make it possible to quickly apply the 5' nuclease assay to any allelic discrimination application where high throughput is of paramount concern.

From: *Methods in Molecular Biology, vol. 212:*
Single Nucleotide Polymorphisms: Methods and Protocols
Edited by: P-Y. Kwok © Humana Press Inc., Totowa, NJ

1.1. 5' Nuclease Assay

In the 5' nuclease PCR assay as first described by Holland et al. *(1,2)*, a hybridization probe included in the PCR is cleaved by the 5' nuclease activity of *Taq* DNA polymerase only if the probe target is being amplified. By using a fluorogenic probe, first synthesized by Lee et al. *(3)*, cleavage of the probe can be detected without post-PCR processing. The fluorogenic probe consists of an oligonucleotide labeled with both a fluorescent reporter dye and a quencher dye. In the intact probe, proximity of the quencher reduces the fluorescence signal observed from the reporter dye due to Förster resonance energy transfer (FRET; *4*). Cleavage of the fluorogenic probe during the PCR assay liberates the reporter dye, causing an increase in fluorescence intensity. This process is diagrammed in **Fig. 1**. The ABI Prism Sequence Detection Systems measures this increase in fluorescence during the thermal cycling of PCR, providing "real-time" detection of PCR product accumulation.

1.2. SNP Detection

Figure 2 diagrams how fluorogenic probes and the 5' nuclease assay can be used for allelic discrimination. For a bi-allelic system, probes specific for each allele are included in the PCR assay. The probes can be distinguished because they are labeled with different fluorescent reporter dyes (FAM™ and VIC™ in **Fig. 2**). A mismatch between probe and target greatly reduces the efficiency of probe hybridization and cleavage. Thus, substantial increase in FAM or VIC fluorescent signal indicates homozygosity for the FAM- or VIC-specific allele. An increase in both signals indicates heterozygosity.

The design of fluorogenic probes has been greatly simplified by the discovery that probes with a reporter dye on the 5' end and a quencher dye on the 3' end exhibit adequate quenching for the probe to perform in the 5' nuclease assay *(5)*. This may seem contradictory to the requirement that reporter and quencher must be in close proximity for quenching by FRET to occur. Nevertheless, it is possible

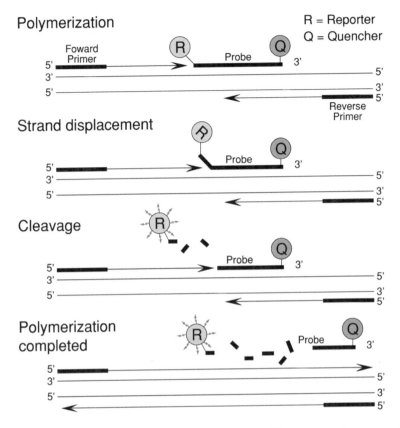

Figure 1. PCR amplification and detection with fluorogenic probes in the 5' nuclease assay. The main steps in the reaction sequence are polymerization, strand displacement, and cleavage. Two dyes, a fluorescent reporter (R) and a quencher (Q), are attached to the fluorogenic probe. When both dyes are attached to the probe, reporter dye emission is quenched. During each extension cycle, the DNA polymerase cleaves the reporter dye from the probe. Once separated from the quencher, the reporter dye fluoresces.

for the dyes to approach each other closely because a single-stranded oligo is flexible enough to bend and contort in solution. These contortions occur quickly compared to the lifetime of the excited state of the reporter, and the quenching observed is a time-resolved average of a population of probes in all bending configurations. There is no

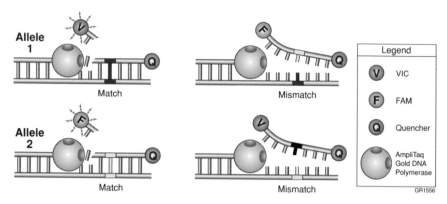

Fig. 2. Allelic discrimination assay design strategy with fluorogenic probes in the 5' nuclease assay. The presence of a mismatch between probe and target destabilizes probe binding during strand displacement, reducing the efficiency of probe cleavage.

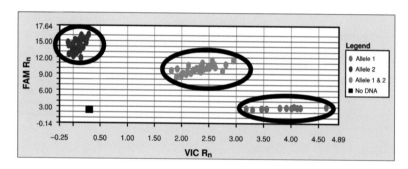

Fig. 3. Discrimination of alleles for the human SNP rs2589 (WIAF 270). Endpoint results using probes VIC-ATGCCC̲CAAGCAG-NFQ-MGB and FAM-TGCCT̲CAAGCAGC-NFQ-MGB on genomic DNA samples from human individuals.

doubt that having dyes at opposite ends of the probe can reduce net quenching and therefore overall signal strength. On the other hand, with both fluorophores close together near the 5' end of the probe, cleavage can occur downstream from both dyes resulting in no signal whatsoever. Thus, the loss of net quenching due to terminal placement of the dyes is more than offset by the fact that any cleavage event will generate signal.

By expanding the region of the probe that can be cleaved to generate signal, it might be expected that terminal placement of the dyes would compromise the ability of a probe to discriminate against mismatches. In fact, Livak et al. *(6)* demonstrated that probes with a reporter at the 5' end and a quencher at the 3' end can discriminate two alleles in the human insulin gene that differ by only a single A-T nucleotide substitution. **Figure 3** shows an example where alleles that differ by an A-G substitution have been distinguished in the DNA from 90 human individuals.

There are a number of factors that contribute to the discrimination based on a single mismatch that is observed in **Fig. 3**. First, there is the thermodynamic contribution caused by the disruptive effect of a mismatch on hybridization. A mismatched probe will have a lower melting temperature (T_m) than a perfectly matched probe. Proper choice of an annealing/extension temperature in the PCR will favor hybridization of an exact-match probe over a mismatched probe. Second, the assay is performed under competitive conditions; that is, both probes are present in the same reaction tube. Therefore, part of the discrimination against a mismatched probe is that it is prevented from binding because stable binding of an exact-match probe blocks hybridization. Third, the 5' end of the probe must start to be displaced before cleavage occurs. The 5' nuclease activity of Taq DNA polymerase actually recognizes a forked structure with a displaced 5' strand of at least 1–3 nucleotides *(7)*. Once a probe starts to be displaced, complete dissociation occurs faster with a mismatch than with an exact match. This means there is less time for cleavage to occur with a mismatched probe. In other words, the presence of a mismatch promotes displacement of the probe rather than cleavage of the probe.

2. Materials

2.1. TaqMan® MGB Probes

As just stated, one of the keys to using the 5' nuclease assay for allelic discrimination is that a mismatched probe has a lower T_m

than a perfectly matched probe. This creates a T_m window between the T_m of the perfectly matched probe and the T_m of the mismatched probe. Discrimination of alleles is achieved by using an annealing/ extension temperature within the T_m window. As probes increase in length, the effect of a single mismatch becomes less disruptive. This means that, for longer probes, the difference in T_m between a matched and mismatched probe is less, leading to a smaller T_m window. Thus, shorter probes should display better mismatch discrimination.

Scientists at Epoch Biosciences have found that the conjugation of a minor groove binder (MGB) to oligonucleotides stabilizes nucleic acid duplexes, causing a dramatic increase in oligonucleotide T_m *(8,9)*. Increases in T_m of as much as 49°C were observed for AT-rich octanucleotides. Fluorogenic probes with MGB attached to the 3' end perform well in the 5' nuclease assay *(10)*. They are an improvement over unmodified probes because shorter sequences (13- to 20-mers) can be used to obtain probes that have an optimal T_m (65–67°C). Thus, attachment of MGB enables the use of shorter fluorogenic probes, which results in improved mismatch discrimination. In a study performed at Applied Biosystems, match and mismatch T_ms were experimentally measured for a set of 60 MGB probes ranging in size from 13–18 nucleotides. The average T_m (match-mismatch) was 9.7°C. This broad T_m window makes it easy to design probes that have a match T_m above the annealing/extension temperature of PCR (nominally 60°C) and a mismatch T_m below this temperature. **Figure 4** diagrams the TaqMan MGB probes now available from Applied Biosystems (P/N 4316032, 4316033, 4316034).

Figure 5 demonstrates the dramatic improvement in assay specificity enabled by MGB probes. Although the MGB probe shown in **Fig. 5** is less than half the length of the conventional probe, the experimentally measured T_m for both probes is approx 65°C. When used in the 5' nuclease assay, the MGB probe shows better performance both in producing higher signal when the probe matches the template and in producing negligible signal when there is a single base mismatch between probe and template.

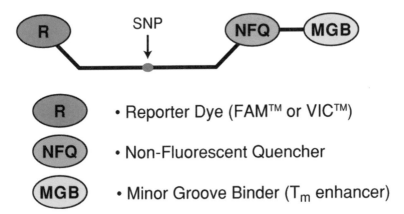

Fig. 4. Diagram of fluorogenic TaqMan® MGB probe.

38-mer ACTTTTCTGTAAGTAGA**T**ATAACTTTTCAAAAAGACAG

17-mer CTGTAAGTAGA**T**ATAAC-MGB

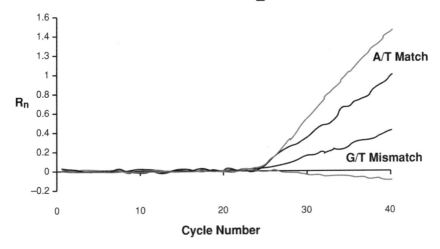

Fig. 5. Results using fluorogenic probes to detect a single base difference in an AT-rich segment of human DNA. The sequence analyzed is a polymorphism in the human thiopurine methyltransferase gene (*TPMT*). The plot shows real-time PCR results for a conventional 38-mer probe (black traces) and a 17-mer MGB probe (red traces) using templates that are either completely homologous to the probes (A/T Match) or have a single base mismatch with the probes (G/T Mismatch).

2.2. TaqMan® Universal PCR Master Mix

Applied Biosystems' vision in developing the 5' nuclease assay was to create a closed tube PCR assay that works the first time and every time. In order to make this a reality, it is necessary to eliminate the optimization of reaction parameters for each individual assay. This has been accomplished by developing specific guidelines for probe and primer selection. If these guidelines are followed, the assay is guaranteed to work using TaqMan Universal PCR Master Mix, universal primer and probe concentrations, and a universal thermal cycling protocol. The most radical aspect of these guidelines is the use of only very small amplicons (50–150 bp). This does not mean larger amplicons will fail, but rather that larger amplicons might require more extensive optimization. Another key aspect of this system is the use of AmpliTaq® Gold DNA polymerase in the TaqMan Universal PCR Master Mix. This enzyme is active only after incubation at elevated temperatures. Using this enzyme provides an invisible hot start to any amplification reaction. This reduces the requirements on primer design because artifacts such as primer dimer are much less of a problem when a Hot Start is used in PCR. Thus, by using AmpliTaq Gold, it is easier to get primers that are going to work the first time.

TaqMan Universal PCR Master Mix (Applied Biosystems P/N 4304437) combines all components required for the 5' nuclease assay except for probes, primers, and sample. Buffer, magnesium chloride, nucleotide triphosphates, and other components have been optimized for robust performance when used in conjunction with the probe and primer selection guidelines detailed in **Subheading 3.1.** Key proprietary components have been included to enhance performance even with difficult GC-rich templates. In addition, the Master Mix contains AmpErase® UNG to provide added protection against carry-over contamination and a passive internal reference to normalize non-PCR-related fluorescence fluctuations. Recommended storage conditions are 2–8°C, so no thawing is required before use.

Table 1
Rules for Selecting Probes

Priority	Guideline
1	Avoid probes with a guanine residue at the 5' end. A guanine residue adjacent to the reporter dye will quench the reporter fluorescence even after cleavage.
2	Select probes with a Primer Express software-estimated T_m of 65–67°C.
3	Make TaqMan MGB probes as short as possible without being shorter than 13 nucleotides.
4	Avoid runs of an identical nucleotide. This is especially true for guanine, where runs of four or more should be avoided.
5	Position the polymorphic site in the central third of the probe. If it is necessary to break this rule, it is preferable to shift the polymorphic site towards the 3' end of the probe because this places the potential mismatch in the minor groove binder domain. Do not place the polymorphic site within the first two or last two nucleotides of the probe.
6	For probes that are over 50% GC, it is preferable to have probes with more C residues than G residues.

3. Methods

3.1. Design of Probes and Primers

User Bulletin "Primer Express Version 1.5 and TaqMan MGB Probes for Allelic Discrimination" (Applied Biosystems P/N 4317594) describes in detail how to design primers and probes for successful allelic discrimination assays. These guidelines make use of version 1.5 of the Primer Express® Oligo Design software in order to estimate the T_ms of primers and MGB probes. The same guidelines and procedures are also used for version 2.0 of the Primer Express Oligo Design software.

3.1.1. Probe Design

Probes are selected first according to the following general rules (*see* **Table 1** and **Note 1**).

Within the Primer Express software, here are the detailed steps for designing the probes:

1. Copy or import the sequence containing the polymorphic site into a "TaqMan MGB Probe & Primer Design" document. If possible, the sequence should extend from approx 100 nucleotides upstream of the polymorphic site to 100 nucleotides downstream. At the polymorphic site, have the sequence corresponding to one of the two alleles (arbitrarily designated Allele 1). It is useful to have the "Double Stranded" checkbox checked so that both strands of the sequence are visible.

2. Select a region containing potential probe sequences.

   a. Highlight a segment of at least 20 nucleotides consisting of the polymorphic site and approximately 10 nucleotides in both the 5' and 3' directions.

   b. From the "Edit" menu, select "Copy."

   c. From the "File" menu, scroll to "New," and select "TaqMan® MGB Probe Test Document." A "TaqMan® MGB Probe Test" document appears.

   d. Click the "Probe 1" text box.

   e. From the "Edit" menu, select "Paste." Primer Express copies the sequence into the "TaqMan MGB Probe Test" document and calculates the T_m using a specialized algorithm for TaqMan MGB probes.

3. Select potential probe sequences in the complementary strand.

   a. Return to the "Sequence" tab in the "TaqMan® MGB Probe" document for Allele 1. The polymorphic sequence and surrounding nucleotides should still be highlighted.

   b. From the "Edit" menu, select "Copy Complement."

   c. Return to the "TaqMan® MGB Probe Test" document and click the "Probe 2" text box.

   d. From the "Edit" menu, select "Paste." Primer Express copies the complementary sequence into the test document and calculates the T_m of the oligonucleotide.

4. Identify a probe that best fits the guidelines outlined in **Table 1**.

   a. Within the TaqMan MGB Probe Test document, highlight potential probe sequences in either the "Probe 1" or "Probe 2" box. Primer Express re-calculates the T_m based on only the highlighted nucleotides. This makes it easy to test different probe sequences in order to identify a probe with an appropriate T_m (65–67°C) that also matches the other guideline criteria.

b. It is important to look for probes in both the Top strand ("Probe 1" box) and Bottom strand (complementary sequence in "Probe 2" box). Because of the asymmetric placement of the minor groove binder at the 3' end, complementary TaqMan MGB probes do not necessarily have the same T_m. Thus, it may be possible to obtain a substantially shorter probe in one strand compared to its complement.

c. Highlight the final selected probe sequence. From the "Edit" menu, select "Trim." Primer Express eliminates all but the selected nucleotide sequence.

d. Copy and paste the final sequence for the Allele 1 probe into a text document for subsequent ordering.

e. If the Allele 1 probe was selected from the Top strand ("Probe 1" box), delete the sequence in the "Probe 2" box. If the Allele 1 probe was selected from the Bottom strand ("Probe 2" box), delete the sequence in the "Probe 1" box.

5. Select the Allele 2 probe.

a. Return to the "Sequence" tab in the "TaqMan MGB Probe" document for Allele 1. The polymorphic sequence and surrounding nucleotides should still be highlighted.

b. If the Allele 1 probe was selected from the Top strand, select "Copy" from the "Edit" menu. If the Allele 1 probe was selected from the Bottom strand, select "Copy Complement" from the "Edit" menu. It is important to perform this step correctly to be sure that the Allele 1 and Allele 2 probes are from the same strand.

c. Return to the "TaqMan® MGB Probe Test" document and click the empty "Probe" text box. From the "Edit" menu, select "Paste."

d. At the polymorphic site, change the sequence to correspond to the sequence for Allele 2. If the "Copy Complement" command was used in **step 5b**, be sure to enter the appropriate complementary Allele 2 sequence.

e. Highlight potential probe sequences in order to identify a probe with an appropriate T_m(65–67°C) that also matches the other guideline criteria. Compared to the Allele 1 probe, the Allele 2 probe may be 1–2 nucleotides shorter or longer, and may be shifted a few nucleotides to the left or right.

Table 2
Rules for Selecting Forward and Reverse Primers

Priority	Guideline
1	Select primers with a Primer Express software-estimated T_m of 58–60°C.
2	Avoid runs of an identical nucleotide. This is especially true for guanine, where runs of four or more should be avoided.
3	Keep the guanine + cytosine content within 20–80%.
4	Place the forward and reverse primers as close as possible to the probes without overlapping the probes.
5	The last five nucleotides at the 3' end should contain no more than two guanine + cytosine residues.

   f. Highlight the final selected probe sequence. From the "Edit" menu, select "Trim." Copy and paste the final sequence for the Allele 2 probe into a text document for subsequent ordering.

3.1.2. Primer Design

After the probes have been designed, the forward and reverse primers are selected according to the following general rules (*see* **Table 2** and **Note 2**).

Within the Primer Express software, here are the detailed steps for designing the primers:

1. Mark the location of the probes in the locus sequence.
   a. Return to the "Sequence" tab in the "TaqMan MGB Probe" document for Allele 1.
   b. Click the "Probe" tool.
   c. Highlight the segment that includes the sequences for the Allele 1 and Allele 2 probes.
2. Automatic selection of primers.
   a. Ensure that the "Limit 3' G+C" checkbox is checked.
   b. Select Find "Primers/Probes Now" from the "Options" menu.
   c. If the software finds acceptable primers:
      i. Click the "Primers" tab. Select a set of primers from the list that will produce the shortest amplicon while satisfying the primer selection guidelines listed in the table above.
      ii. Copy and paste the final primer sequences into a text document for subsequent ordering.

    d. If the software does not find acceptable primers, proceed to manual selection of primers.

3. Manual selection of primers.

    a. Copy potential primer sequences to a "Primer Test Document."

       i. Highlight the 40 nucleotides upstream of the marked Probe region. (The Primer Test document only accepts sequences that are 40 nucleotides or shorter.)

       ii. From the "Edit" menu, select "Copy."

      iii. From the "File" menu, scroll to "New," and select "Primer Test Document." A "Primer Test" document appears. Do not change the default "Primer Concentration" and "Salt" conditions. Also, do not check the "ppT" checkbox.

      iv. Click the "Forward Primer" text box.

       v. From the "Edit" menu, select "Paste." Primer Express copies the sequence into the Primer Test document and calculates the T_m of the oligonucleotide.

      vi. Return to the "Sequence" tab in the "TaqMan MGB Probe" document for Allele 1. Highlight the 40 nucleotides downstream of the marked Probe region.

     vii. From the "Edit" menu, select "Copy Complement."

    viii. Return to the Primer Test document and click the "Reverse Primer" text box.

      ix. From the "Edit" menu, select "Paste." Primer Express copies the complementary sequence into the test document and calculates the T_m of the oligonucleotide.

    b. Identify primers that best fit the primer selection guidelines outlined in the table above.

       i. Highlight potential primer sequences in the "Forward Primer" box. Primer Express re-calculates the T_m based on only the highlighted nucleotides. Identify a sequence that has an appropriate T_m (58–60°C) and best fits the other criteria listed in the table.

      ii. With the final sequence highlighted, select "Trim" from the "Edit" menu. Copy and paste the sequence for the Forward Primer into a text document for subsequent ordering.

      iii. Highlight potential primer sequences in the "Reverse Primer" box. Identify a sequence that has an appropriate T_m (58–60°C) and best fits the other criteria listed in the table.

      iv. With the final sequence highlighted, select "Trim" from the "Edit" menu. Copy and paste the sequence for the Reverse Primer into a text document for subsequent ordering.

3.2. Protocol for 25 μL Reactions

By following the probe/primer selection procedure and using TaqMan Universal PCR Master Mix, reactions can be run reliably using fixed concentrations of 200 nM each probe and 900 nM each primer. It is convenient to pre-mix the probes and primers as a 10X Probe/Primer Mix: 2 μM VIC Probe, 2 μM FAM Probe, 9 μM Forward Primer, 9 μM Reverse Primer, 10 mM Tris-HCl, pH 8.0, 1 mM EDTA.

With storage at −20°C, this mix should be stable for at least 1 yr. Just before use, the mix should be thawed with minimal exposure to light, especially direct sunlight. After thawing, the mix should be vigorously vortexed before pipetting into the reaction mix. Promptly return the 10X Probe/Primer mix to the −20°C freezer after use.

For reactions run in 96-well plates, the reaction volume is typically 25 μL. For 96 samples, the following Reaction Mix is prepared: 1300 μL 2X TaqMan Universal PCR Master Mix, 260 μL 10X Probe/Primer Mix, 520 μL H$_2$O.

Samples are added to the plate by pipetting 5 μL genomic DNA (at a concentration of 1–10 ng/μL) into each well. Then, 20 μL of the Reaction Mix are added to each well using a repeating pipettor. The plate is then sealed using an Optical Adhesive Cover (ABI PRISM™ Optical Adhesive Starter Kit, P/N 4313663). Prepared plates can be stored at 4°C for up to 72 h.

For thermal cycling, the plate can be placed in any of the following instruments: GeneAmp® PCR System 9600, GeneAmp® PCR System 9700, ABI PRISM® 7700 Sequence Detection System, or ABI PRISM® 7900HT Sequence Detection System. For all instruments except the ABI PRISM® 7900HT, it is important to use the Compression Pad that comes with the Optical Adhesive Cover on top of the plate. Thermal cycling parameters are shown in **Table 3**.

If the amount of genomic DNA is 10 ng or less, the number of cycles should be increased to 40.

After thermal cycling, the fluorescence in the plate can be measured using either the ABI PRISM 7700 or 7900HT. The plate should be set up as an Allelic Discrimination plate following the instruc-

Table 3
Thermal Cycling Parameters for 25 μL Reactions

	AmpErase UNG Activation Hold	AmpliTaq Gold Enzyme Activation Hold	PCR Cycle (35 cycles)	
			Denature	Anneal/extend
Temp.	50°C	95°C	92°C	60°C
Time	2 min	10 min	15 s	1 min

tions in the instrument manual. The SDS software will display the fluorescence results as a scatter plot of VIC signal versus FAM signal. Up to four distinct clusters should be observed. Using the software tools, each cluster can be selected and marked as Allele 1 Homozygote, Heterozygote 1/2, Allele 2 Homozygote, or No Amp (No DNA). *See* **Note 3** for trouble shooting when the performance of the assay is not robust.

3.3. Protocol for 5 μL Reactions

3.3.1. Protocol with Standard DNA Samples

The ABI PRISM 7900HT enables fluorescence detection in 384-well plates. In 384-well plates, the reaction volume can be reduced to as little as 5 μL. For 384 samples, the following Reaction Mix is prepared: 1050 μL 2X TaqMan Universal PCR Master Mix, 210 μL 10X Probe/Primer Mix. Samples are added to the plate by pipetting 2 μL genomic DNA (at a concentration of 1–10 ng/μL) into each well. Then, 3 μL of the Reaction Mix are added to each well using a repeating pipettor. The plate is then sealed using an Optical Adhesive Cover (ABI PRISM™ Optical Adhesive Starter Kit, P/N 4313663). Prepared plates can be stored at 4°C for up to 72 h.

Thermal cycling can be performed in either the GeneAmp PCR System 9700 or ABI PRISM 7900HT Sequence Detection System. Using the GeneAmp 9700, it is important to put the Compression Pad that comes with the Optical Adhesive Cover on top of the plate. Thermal cycling parameters are shown in **Table 4**.

Table 4
Thermal Cycling Parameters for 5 µL Reactions

	AmpErase UNG Activation Hold	AmpliTaq Gold Enzyme Activation Hold	PCR Cycle (35 cycles)	
			Denature	Anneal/extend
Temp.	50°C	95°C	92°C	60°C
Time	2 min	10 min	15 s	1 min

If the amount of genomic DNA is 4 ng or less, the number of cycles should be increased to 40.

After thermal cycling, the fluorescence in the plate is measured using the ABI PRISM 7900HT. The plate should be set up as an Allelic Discrimination plate following the instructions in the instrument manual. The SDS software will display the fluorescence results as a scatter plot of VIC signal versus FAM signal. Up to four distinct clusters should be observed. Using the software tools, each cluster can be selected and marked as Allele 1 Homozygote, Heterozygote 1/2, Allele 2 Homozygote, or No Amp (No DNA).

3.3.2. Protocol with Dried Down DNA Samples

Pipetting volumes less than 5 µL can sometimes lead to reduced precision. This is especially true for some robotic reagent dispensing systems. An alternative procedure for performing reactions in 384-well plates is to dispense 5 µL genomic DNA (at a concentration of 0.4–4 ng/µL) into each well. These DNA samples are dried down in the plate by allowing the plate to sit open overnight at room temperature. For 384 samples, the following Reaction Mix is prepared: 1050 µL 2X TaqMan Universal PCR Master Mix, 210 µL 10X Probe/Primer Mix, 840 µL H₂O.

After dispensing 5 µL of this Reaction Mix to each of the dried down DNA samples, the plate is sealed using an Optical Adhesive Cover. Prepared plates can be stored at 4°C for up to 72 h. The plate is then subjected to thermal cycling and analyzed as described in **Subheading 3.3.1.**

4. Notes

1. For SNP detection, the location of the polymorphism dictates the placement of the probes. Therefore, it is sometimes necessary to use probes that do not meet all the criteria listed in **Table 1** in **Subheading 3.1.1.** This is why the guidelines are listed in order of priority. Rules of lower priority should be broken first. Thus, using a short probe (priority 3) is more important than having a probe with more Cs than Gs (priority 6). For example, a 13-mer probe with 5 Gs and 3 Cs is preferable to a 16-mer probe with 6 Cs and 2 Gs. As outlined in the table, shifting the polymorphic site outside the central third of the probe (priority 5) can also be used if this results in a shorter probe (priority 3) or avoids a run of identical nucleotides (priority 4). In balancing these criteria, it is important to consider probes from either strand. In general, the goal is to obtain the shortest possible probe (but not shorter than 13 nucleotides) that best fits the other criteria.

2. Similar to probes, **Table 2** in **Subheading 3.1.2.** lists primer selection criteria in order of priority. The most restrictive criterion is the priority 5 rule that states the last five nucleotides at the 3' end should contain no more than two G + C residues. This rule is designed to reduce duplex stability at the 3' end of the primer in order to minimize mis-priming at non-specific sites. This guideline can be broken if it results in a substantially shorter amplicon (priority 4). If the priority 5 rule is broken, a primer with three G + Cs at the 3' end is preferable to a primer with four G + Cs at the 3' end. In practice, a primer where all five nucleotides at the 3' end are G + C can almost always be avoided.

3. Assays designed and performed as described here should perform well without optimization. If poor performance is encountered, factors other than assay design need to be considered. For example, primers or probes derived from erroneous sequence information cannot be expected to generate signal. Another possibility is that the situation in the genome is more complex than originally assumed. The presence of unidentified polymorphisms that affect primer or probe binding can negatively impact performance. This can result in an assay that performs well in some individual samples but not in others. Other assays have been performed where it appears that every sample is heterozygous. This indicates cases where a duplication in the genome has been mis-classified as a SNP. In general, repeated sequences should be avoided in designing primers and probes.

Artificial templates provide an important control that can be used to evaluate assay performance. Artificial templates are chemically synthesized oligonucleotides that contain the primer and probe regions of the target amplicon. Separate artificial templates can be synthesized for each allele. If poor signal is observed with artificial templates, this indicates a problem with the primer and probe reagents. One possibility is that the wrong reagents were mixed in preparing the 10X Probe/Primer mix. An error due to mis-identification of primers or probes or due to incorrect mixing will lead to a failed assay. It is also possible that a mistake was made in the synthesis of one of the primers or probes. After checking to make sure the primer and probe sequences are correct, re-synthesis of the primers and/or probes should solve the problem. Good signal observed with the artificial templates verifies the integrity of the primers and probes and the validity of the assay design. If poor signal is then observed with genomic DNA, factors such as erroneous sequence information, unidentified polymorphisms, or other genome complexities need to be considered.

If it is desired to re-design an assay, it is usually only necessary to re-design the primers. This is because it is the primers that are responsible for the exponential amplification of the 5' nuclease assay. Therefore, the primers determine what sequence is amplified and the efficiency of that amplification. If the appropriate sequence is amplified, the use of TaqMan MGB probes ensures that the sequence will be detected with single base discrimination. If new primers are synthesized, they should be tested in all pairwise combinations in order to determine which pair gives the best results.

References

1. Holland, P. M., Abramson, R. D., Watson, R., and Gelfand, D. H. (1991) Detection of specific polymerase chain reaction product by utilizing the 5' in place of 3' exonuclease activity of *Thermus aquaticus* DNA polymerase. *Proc. Natl. Acad. Sci. USA* **88**, 7276–7280.
2. Holland, P. M., Abramson, R. D., Watson, R., Will, S., Saiki, R. K., and Gelfand, D. H. (1992) Detection of specific polymerase chain reaction product by utilizing the 5' in place of 3' exonuclease activity of *Thermus aquaticus* DNA polymerase. *Clin. Chem.* **38**, 462–463.
3. Lee, L. G., Connell, C. R., and Bloch, W. (1993) Allelic discrimination by nick-translation PCR with fluorogenic probes. *Nucleic Acids Res.* **21**, 3761–3766.

4. Förster, V. T. (1948) Zwischenmolekulare energiewanderung und fluoresenz. *Ann. Physics* (Leipzig) **2**, 55–75.

5. Livak, K. J., Flood, S. A. J., Marmaro, J., Giusti, W., and Deetz, K. (1995) Oligonucleotides with fluorescent dyes at opposite ends provide a quenched probe system useful for detecting PCR product and nucleic acid hybridization. *PCR Methods Appl.* **5**, 357–362.

6. Livak, K. J., Marmaro, J., and Todd, J. A. (1995) Towards fully automated genome-wide polymorphism screening. *Nat. Genet.* **9**, 341–342.

7. Lyamichev, V., Brow, M. A. D., and Dahlberg, J. E. (1993) Structure-specific endonucleolytic cleavage of nucleic acids by eubacterial DNA polymerases. *Science* **260**, 778–783.

8. Afonina, I., Zivarts, M., Kutyavin, I., Lukhtanov, E., Gamper, H., and Meyer, R. B. (1997) Efficient priming of PCR with short oligonucleotides conjugated to a minor groove binder. *Nucleic Acids Res.* **25**, 2657–2660.

9. Kutyavin, I. V., Lukhtanov, E. A., Gamper, H. B., and Meyer, R. B. (1997) Oligonucleotides with conjugated dihydropyrroloindole tripeptides: Base composition and backbone effects on hybridization. *Nucleic Acids Res.* **25**, 3718–3723.

10. Kutyavin, I. V., Afonina, I. A., Mills, A., Gorn, V. V., Lukhtanov, E. A., Belousov, E. S. , et al. (2000) 3'-Minor groove binder-DNA probes increase sequence specificity at PCR extension temperatures. *Nucleic Acids Res.* **28**, 655–661.

10

Genotyping SNPs by Minisequencing Primer Extension Using Oligonucleotide Microarrays

Katarina Lindroos, Ulrika Liljedahl, and Ann-Christine Syvänen

1. Introduction

A promising approach towards high-throughput genotyping of single nucleotide polymorphisms (SNPs) is to use arrays of immobilized oligonucleotides in miniaturized assays *(1,2)*. A significant advantage of performing the assays in microarray formats is that the costs of genotyping are reduced because many SNPs are analyzed simultaneously in each sample, and because the reaction volumes employed on the microarrays are small. The three major reaction principles that are currently in use for genotyping SNPs, namely hybridization with allele-specific oligonucleotide (ASO) probes *(3)*, oligonucleotide ligation *(4)* or DNA polymerase-assisted primer extension *(5)* have all been utilized in microarray-based assay formats. Simultaneous genotyping of multiple SNPs by ASO hybridization is hampered by the poor specificity of ASO probes to discriminate between SNP genotypes in large, diploid genomes *(6,7)*. Therefore the enzyme-assisted methods are gaining accep-

From: *Methods in Molecular Biology, vol. 212:*
Single Nucleotide Polymorphisms: Methods and Protocols
Edited by: P-Y. Kwok © Humana Press Inc., Totowa, NJ

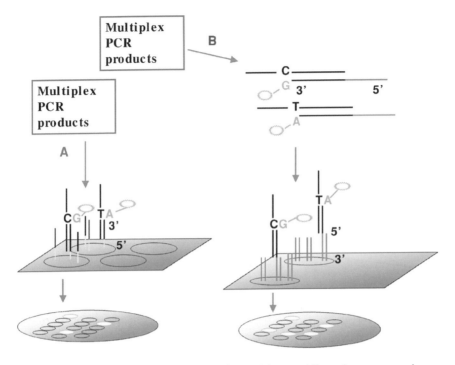

Fig. 1. Principle of minisequencing with specific primers on micro-arrays **(A)** and using "Tag-arrays" **(B)**.

tance as more specific alternatives than ASO hybridization for multiplex SNP typing *(8,9)*.

We have developed a microarray system based on "minisequencing" single nucleotide primer extension that allows highly specific, multiplex genotyping of SNPs *(5)*. In the minisequencing reaction a DNA-polymerase is used to extend detection primers that anneal immediately adjacent to the sites of the SNPs with labeled nucleotide analogues that are complementary to the nucleotide(s) at the SNP sites *(10)* (*see* **Fig. 1A**). In the microarray format of the method, detection primers specific for the SNPs to be analyzed are attached covalently to surface-activated glass microscope slides through an amino group in their 5'-end. The primers are applied to the micro-scope slide using an array spotter in a configuration of 80 "subarrays" that may contain up to 240 primers each *(11)* (*see* **Fig. 2**). This for-

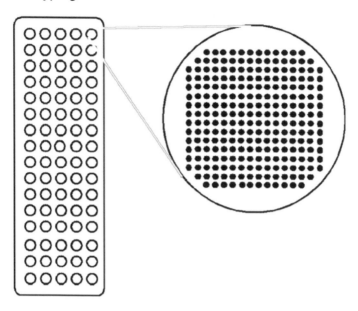

Fig. 2. A microscope glass slide with 80 "subarrays" of 4.3 mm in diameter. Each subarray may contain up to 240 oligonucleotide spots of ~100 µm in diameter at a spot to spot distance of 150 µm. The capacity is thus 19,200 spots (genotypes) per slide.

mat allows the generation of several thousand genoptypes per slide. The throughput of our genotyping method is limited mainly by the requirement of establishing and performing multiplex PCRs of the DNA regions spanning the SNPs before the genotyping reactions on the microarrays. The PCR products spanning the SNPs are allowed to anneal to the immobilized detection primers followed by extension of the primers with fluorescent ddNTPs using a DNA polymerase (*see* **Fig. 1A**). In the protocol presented below, ddNTPs labeled with a single fluorophore (TAMRA) are used to detect all four nucleotides in separate reactions for each sample *(12)*. After the minisequencing reactions, the incorporated fluorescence is measured in a fluorescence scanner. **Figure 3** shows an example of the result from genotyping one sample for 50 SNPs.

We also present a protocol for a modification of the method, in which cyclic minisequencing reactions with TAMRA-labeled

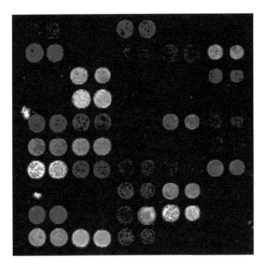

Fig. 3. Example of a result from minisequencing one sample for 50 SNPs with specific primers on the microarray with Tamra-labeled dd NTPs. The oligonucleotides are printed as duplicates on the slide. (Unpublished data by Liljedahl et al.)

ddNTPs are performed in solution using primers with 5'-"Tag" sequences, followed by capture of the reaction products on microarrays carrying oligonucleotides that are complementary to the "Tag" sequences ("cTags") (*see* **Fig. 1B**). The concept of using "tagged" polymerase chain reaction (PCR) primers was first described for analyzing gene expression in yeast by PCR *(13)*, and has later been applied to SNP genotyping by primer extension and capture on fluorescent microparticles *(14)*, high-density oligonucleotide arrays (Affymetrix, GenFlex[TM] arrays; *[15]*), and medium-density, custom made oligonucleotide arrays *(16)*. Our modification of the "Tag-array" method makes use of the format with 80 subarrays per microscope slide, to which the cTags are immbobilized through a 3'- amino group as described above. **Figure 4** compares the steps of the two procedures, for which the detailed protocols are given below. All materials and equipment required for setting up and performing the assay are generally available from common suppliers.

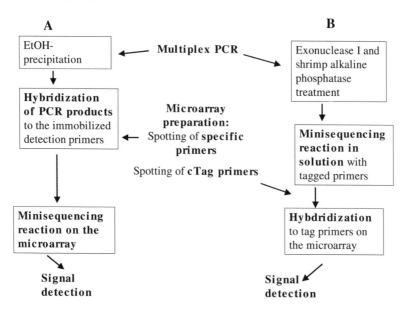

Fig. 4. Flow sheet of the genotyping procedures with specific primers on microarrays (**A**) and using "Tag-arrays" (**B**).

Thus the technology can be established in molecular biology laboratories wishing to increase their "in house" throughput for genotyping SNPs.

2. Material

2.1. Reagents

2.1.1. Multiplex PCR

1. Thermostable DNA polymerase AmpliTaq Gold® 5 U/µL (Applied Biosystems, Foster City, CA).
2. 10X concentrated PCR Buffer (GeneAmp® 10X PCR Buffer, Applied Biosystems): 100 mM Tris-HCl, pH 8.3, 500 mM KCl, 15 mM MgCl$_2$, 0.01% (w/v) gelatin.
3. dNTP mixture: 2 mM dATP, 2 mM dCTP, 2 mM dGTP, and 2 mM dTTP.

2.1.2. Microarray Preparation (see **Note 1**)

1. Microscope glass slides (Menzel-Gläser, Braunschweig, Germany).
2. Alconox® detergent powder (Aldrich).
3. 99.5% EtOH.
4. Aminopropylsilane-solution: 1% aminopropylsilane in 95% acetone/ H_2O. 150 mL acetone, 8 mL H_2O, 1.6 mL 3-aminoporopyl-triethoxysilane (Aldrich). Prepare prior to use in a hood.
5. Acetone
6. 0.2% *p*-Phenylenediisothiocyanate-solution prepared prior to use in a hood. Add the reagents in the following order: 300 mg 1,4-phenyl-enediisothiocyanate (Fluka), 144 mL *N,N*-dimethylformamide (Merck), 16 mL pyridine (Merck).
7. Methanol.
8. For the minisequencing reaction with specific primers (*see* **Subheading 3.2.1.** below): 5' amino-modified oligonucleotide primers with a spacer sequence of 15T-residues in their 5' ends at a concentration of 25 μ*M* in 400 m*M* sodium carbonate/bicarbonate buffer, pH 9.0. For the tag-arrays: 3' amino-modified oligonucleotide primers with a spacer sequence of 15T-residues in their 3' ends at a concentration of 25 μ*M* in 400 m*M* sodium carbonate/bicarbonate buffer, pH 9.0. The diluted oligonucleotides can be stored at −20°C.
9. 25% NH_4OH.

2.1.3. Minisequencing Reactions with Specific Primers on Microarrays

1. 95–99.5% EtOH.
2. 3 *M* Sodium-acetate, pH 4.8.
3. 70% EtOH at −20°C.
4. 5X Annealing buffer: 1 *M* NaCl, 50 m*M* Tris-HCl, pH 8.0, 5 m*M* EDTA. Store at about 20°C. Prepare 10 mL, which is enough for several minisequencing reactions.
5. Washing solution: 5 m*M* Tris-HCl, pH 8.0, 0.5 m*M* EDTA, 100 m*M* NaCl, 0.1% Triton-X® 100. Fifty milliliters is enough for washing two slides. The solution can be stored for several months at about 20°C.
6. TAMRA-labeled ddNTPs at a 5 μ*M* concentration (NEN™ Life Science Products, Brussels, Belgium). Store the stock fluorescent ddNTPs at 100 μ*M* unexposed to light as aliquots at −20°C.

7. Unlabeled ddNTPs at 10 μM concentration.
8. DNA polymerase, 0.5 U/μL (DynaSeq from Finnzymes OY, Helsinki, Finland or ThermoSequenase AP Biotech, Uppsala, Sweden).
9. 10X reaction buffer: 260 mM Tris-HCl, pH 9.5, 65 mM MgCl$_2$, 2% Triton® X-100.
10. 50 mM NaOH (make fresh every week).
11. 0.2X SSC + 0.1% SDS solution (1X SSC is 15 mM sodium citrate and 150 mM NaCl, pH 7.0). One hundred milliliters is enough for two slides. The solution can be stored for several months at about 20°C.

2.1.4. Reagents for Minisequencing Reactions Using "Tag-Arrays"

1. 25 mM MgCl$_2$.
2. Tris-HCl, 1 M, pH 9.5.
3. Exonuclease I 10 U/μL (USB Corporation, Cleveland, OH).
4. Shrimp alkaline phosphatase 1 U/μL (USB Corporation).
5. DNA polymerase, 0.5 U/μL (DynaSeq or ThermoSequenase).
6. Minisequencing primers with 5' tag sequences that are complementary to the tag-oligonucleotides on the microarray (*see* **Subheading 3.2.1.** below).
7. 10X reaction mix: 15 nM of each tagged minisequencing primer, 0.25 mM of the three unlabeled ddNTPs, 1% Triton-X100.
8. TAMRA-labeled ddNTPs at a 5 μM concentration (NEN™ Life Science Products, Brussels, Belgium). Store the stock fluorescent ddNTPs at 100 μM unexposed to light as aliquots at −20°C.
9. 20X SSC.
10. Hybridization control oligonucleotide, 45 nM, complementary to one of the spotted tag-oligonucleotides on the microarray.
11. SSC for washing.

2.2. Equipment

1. Programmable heat block and facilities to avoid contamination during PCR.
2. Access to oligonucleotide synthesis.
3. Access to microarray manufacturing. We use a custom built robot with TeleChem CMP2 (Sunnyvale, CA) printing pins that produce oligonucleotide spots of 125–150 μm in diameter.

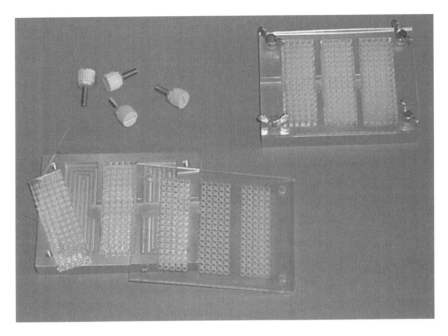

Fig. 5. The microarray reaction rack.

4. Hybridization oven or incubator at 37, 40, 42, 55, and 65°C.
5. Minisequencing reaction rack with a plexyglass cover. We use a custom-made aluminium rack that holds three slides and is covered by a reusable silicon rubber grid (*see* **Subheading 2.2.1.**), which is applied on the arrays to form 80 separate reaction wells on each slide. The plexyglass should have holes to match the wells that are formed with the silicon rubber grid (*see* **Fig. 5**).
6. Shaker at about 20°C, at 40°C, and at 65°C.
7. Centrifuge for microscope slides and Eppendorf tubes.
8. Array scanner and a software for signal analysis. (We use a ScanArray® 5000 with the QuantArray® analysis software, GSI Lumonics, Watertown, MA.)

2.2.1. Preparation of Rubber Grids

Reusable silicon rubber grids are prepared using an inverted 384-well microtiter plate with v-shaped wells (Biometra, Göttingen, Germany) as a mold *(11)*. Liquid silicon rubber Elastosil RT 625A

(Wacker-Chemie GmbH, Munich, Germany) is mixed with Elastosil RT 625B in a 9:1 proportion, and poured into the mold, leaving about 1–2 mm of the tip of the wells uncovered. A total amount of 50 g is enough for one 384-well microtiter plate. The silicon is allowed to harden for at least 24 h at room temperature before it is cut to match the size of the microarray slides.

3. Methods

3.1. Comment on Multiplex PCR

Multiplex PCR amplification of more than 10 fragments reproducibly and successfully from multiple samples has proven to be difficult *(3,11)*. A "touchdown" PCR procedure *(17)* to circumvent differences in amplification efficiency due to differences in melting temperatures between the PCR primers in combination with universal 5'-sequences on the primers may be used *(18)* to unify the reaction kinetics of the primer annealing. To avoid complementary 3' overlap of the primers they can be designed with common terminal dinucleotides, for instance AC, at their 3' end *(19)*. The optimal primer concentration during multiplex PCR may vary from 0.1 to 1.2 μM. Fifty to 500 ng of DNA and 3.5 U of DNA polymerase are standard amounts in 100 μL multiplex PCR reaction. A touchdown PCR procedure that functions well for us is: Initial activation of the polymerase at 95°C for 11 min, then 95°C for 30 s, 65°C–1°C per cycle for 4 min for 5 cycles; 95°C for 30 s, 60°C–0.5°C per cycle for 2 min and 68°C for 2 min for 15 cycles; 95°C for 30 s, 53°C for 30 s, 68°C for 2 min for 14 cycles; 68°C for 2 min *(12)*.

3.2. Minisequencing

3.2.1. Design of Minisequencing Primers

The specific minisequencing primer should hybridize to the DNA template with its 3' end with the nucleotide adjacent to the variant nucleotide to be analyzed. The primers should be approx 20 bases long, all with similar T_m-values close to 57°C (excluding the 15 T-residue). To avoid misincorporation of ddNTPs the primer should not form a 3' hairpin-loop or other complementary structure. Such structures may be avoided by using a primer from the complementary DNA strand. The minisequencing primers to be immobilized contain 15T-residues as spacers 5' of the specific sequence and a 5'-amino group for attachment to the glass surface (*see* **Subheading 3.2.3.**). Minisequencing primers to be used in solution in the tag-array assay are designed with a 5'-tag sequence (*see* **Subheading 3.2.4.**). The tag-oligonucleotide sequences should be around 20–27 bases long, and their T_m should be approx 55°C. Their self similarity and similarity to each other should be low *(16)*. The complementary tag sequences to be immobilized contain 15T-residues as spacers 3' of the specific sequence and a 3'-amino group (*see* **Fig. 1**).

3.2.2. Preparation of Microarrays (see **Note 1**)

The slides are amino-silanized and coated using a bifunctional crosslinker to obtain a isothiocyanate surface that attaches NH_2-modified oligonucleotides *(20)*.

1. Wash the microscope glass slides with approx 1% Alconox® detergent powder in warm tap water, rinse with dH_2O and with 99.5% EtOH and leave to dry at room temperature.
2. Dry the slides at 80°C for 10 min.
3. Incubate the slides in a closed container for 2 h at about 20°C in the aminopropylsilane-solution.
4. Rinse the slides 3 times with acetone, dry at room temperature and at 110°C for 45 min (*see* **Note 2**).
5. Incubate the slides in a closed container for 2 h at room temperature in the 0.2% p-phenylenediisothiocyanate-solution, immerse into a methanol-bath and finally in an acetone-bath.
6. Leave the slides to dry at room temperature in a fume hood and store at 4°C prior to spotting for up to 2 mo.
7. Print the amino-modified oligonucleotides (25 μ*M*) in 400 m*M* sodium carbonate/bicarbonate buffer, pH 9.0 onto the activated slides. Immediately after spotting, expose the slides to vaporized 25% ammonia for 1 h in a closed container.
8. Wash the arrays three times in distilled water. The arrays should be stored at 10°C until use.

3.2.3. Minisequencing Using Specific Primer Arrays

3.2.3.1. ETOH-PRECIPITATION

1. Precipitate the amplified DNA samples by adding 250 μL 95–99.5% EtOH and 10 μL of sodium-acetate (3 M, pH 4.8) to 100 μL of the PCR products (*see* **Note 3**).
2. Leave the samples at −20°C for at least 20 min and centrifuge at 13,000 rpm for 15 min.
3. Discard the supernatant gently and add 1 mL of −20°C 70% EtOH to the samples, follow by a 15 min centrifugation.
4. Discard the EtOH and allow the DNA pellets to dry for 15 min.
5. Dissolved in 40 μL of H_2O (*see* **Note 3**).

3.2.3.2. ANNEALING

1. Denature the amplified and precipitated DNA sample at 95°C for 3 min, quench, and keep on ice.
2. Add 10 μL of 5X annealing buffer to the tube containing the amplified DNA sample in 40 μL of H_2O.
3. Include an annealing mix without DNA to serve as a control by combining 10 μL of annealing buffer and 40 μL H_2O in each assay (*see* **Note 4**).
4. Place the slide with the arrays of specific primers in the reaction rack with the rubber grid and the plexyglass cover and preheat the assembly to 37°C.
5. Pipet 10 μL of each sample or control into four parallel wells on the slide.
6. Place the rack into a humid chamber previously preheated to 37°C and allow the annealing reaction to proceed for 40 min at 37°C.
7. Remove the slide from the rack, and rinse briefly in washing solution. Let the slide dry before reassembling the rack.

3.2.3.3. MINISEQUENCING REACTIONS ON ARRAYS

Because this protocol is for a one-color based detection system, four different reaction mixtures each of which contain one of the four TAMRA-labeled ddNTPs in four reaction wells for each sample are required. Each minisequencing reaction mixture contains one of the four TAMRA-labeled ddATP, ddCTP, ddGTP,

ddTTP analogues at a 0.75 μ*M* concentration together with the other
three unlabeled ddNTPs at a concentration of 0.5 μ*M* in 20 μL of
reaction buffer, containing 1.5 U of DNA polymerase.

1. Mix 9 μL of dH$_2$O with 2 μL of 10X reaction buffer, 1 μL of the
   three unlabeled ddNTPs at 10 μ*M* concentration, 3 μL of one of the
   four TAMRA-labeled ddNTP analogues at 5 μ*M* concentration, and
   3 μL of DNA-polymerase at 0.5 U/μL.
2. Preheat the rack containing the slide to 55°C and add 20 μL of the
   four reaction mixtures to four separate reaction wells for each
   sample.
3. Incubate at 55°C for 15 min in a humid chamber (*see* **Note 5**).
4. Wash the slides using a shaker, twice with dH$_2$O, once with 50 m*M*
   NaOH for 2 min and again with dH$_2$O.
5. Wash the slides using a shaker, twice for 5 min at 65°C in 0.2X SSC
   + 0.1% SDS and finally with dH$_2$O (*see* **Note 6**).
6. Leave the slides to dry in a dark place at room temperature or dry
   them by centrifugation.

3.2.4. Minisequencing Using Generic Tag Arrays

3.2.4.1. EXONUCLEASE I AND SHRIMP ALKALINE PHOSPHATASE TREATMENT

Exonuclease I and shrimp alkaline phosphatase are used to
remove the excess of dNTPs and primers from the PCR. The protocol is given for a one-color based detection system that requires
four different reaction mixtures each of which contains one of the
four TAMRA-labeled ddNTPs, in four separate wells for each
sample. The given amounts are for one reaction well only:

1. Add 1.5 μL of 25 m*M* MgCl$_2$, 0.5 μL of 1 *M* Tris-HCl, pH 9.5,
   0.5 μL of exonuclease I (10 U/μL) and 1 μL of shrimp alkaline phosphatase (1 U/μL) to 7 μL of the multiplex PCR product.
2. Incubate the mixture for 60 min at 37°C and inactivate the enzymes
   for 15 min at 99°C.

3.2.4.2. THE MINISEQUENCING REACTIONS

1. Combine 1.5 μL of the 10X reaction mixture containing the tagged
   minisequencing primers, the three unlabeled ddNTPs and Triton

X-100, 2 µL of DNA polymerase (0.5 U/µL) and 2 µL of one 5 µ*M* TAMRA-labeled ddNTP to 10.5 µL of the exonuclease I and shrimp alkaline phosphatase-treated PCR product.
2. Heat the minisequencing reaction mixtures at 96°C for 3 min, followed by thermocycling them 99 times at 95°C for 20 s and 55°C for 20 s.

3.2.4.3. HYBRIDIZATION TO TAG OLIGONUCLEOTIDES ON THE MICROARRAY

1. Place the slide with the tag oligonucleotide arrays in the reaction rack with the rubber grid and the plexyglass cover and preheat the assembly to 42°C.
2. Add 6.5 µL of 20X SSC and 0.5 µL of 45 n*M* hybridization control oligonucleotide to 15 µL of the minisequencing product.
3. Pipet the mixtures into the reaction wells on the slide.
4. Place the rack into a humid chamber previously preheated to 42°C and allow the hybridization reaction to proceed for 2–4 h at 42°C (*see* **Note 5**). Remove the slide from the rack and rinse briefly with SSC.
5. Wash twice for 5 min in with SSC at 40°C using a shaker and avoiding exposure to light.
6. Leave the slides to dry in a dark place at room temperature or dry them by centrifugation.

3.2.5. Data Interpretation

1. Read the fluorescence intensity signals of each spot using an array scanner.
2. Correct the signal intensity from each spot for the average background in each well by measuring the average background from 5–10 spots immediately below the array, and correcting the signal by subtraction of this average background signal.
3. The genotypes are assigned by calculating the ratio between the corrected signal intensities from the reaction for one of the alleles divided by the signal from the other allele. **Table 1** gives an example of a genotyping result of 6 SNPs from 2 samples.

4. Notes

1. Commercially available slides can be used instead of in-house activated ones. We recommend for example 3D-Link™-slides from SurModics, Inc. (Eden Prairie, MN) that attach NH_2-modified oligo-

**Table 1
Example of Numeric Genotyping Result for 6 SNPs in Two
Samples**

SNP	Variation	Genotype	Sample	R-value[c]
AGTR1 1166 co[a]	A/C	Homozygous	1	0,00010
AGTR1 1166 co		Homozygous	2	180
AGTR1 1166 nc	T/G	Homozygous	1	0,040
AGTR1 1166 nc		Homozygous	2	130
AGTR1 1062 co[a]	A/G	Homozygous	1	5,1
AGTR1 1062 co		Heterozygous	2	1,8
AGTR1 1062 nc	T/C	Homozygous	1	250
AGTR1 1062 nc		Heterozygous	2	1,3
ENOS co[b]	T/G	Homozygous	1	0,0030
ENOS co		Heterozygous	2	2,8
ENOS nc	A/C	Homozygous	1	0,080
ENOS nc		Heterozygous	2	3,9

[a]The SNPs are from the human angiotensin II type 1 receptor gene (GeneBank accession number AF245699). Coding strand of the DNA is denoted co and the noncoding strand is denoted nc.

[b]SNPs are from the human endothelial nitric oxide synthase (GeneBank accession number X76307).

[c]Unpublished results by Lindroos et al.

    nucleotides to their surface. In this case the protocol provided by the manufacturer should be used for spotting the primers on the microarray.

2. The silanization procedure, and the subsequent isothiocyanate-treatment are performed in a fume hood due to the toxicity of the reagents.

3. It has proven to be difficult to amplify more than 10 SNPs per multiplex PCR reaction reproducibly *(11,12)* If the SNPs have been amplified in several multiplex PCRs, the products from each sample can be combined and precipitated together. The amount of 99.5% EtOH should be 2.5 times the total amount of the PCR products and the amount of sodium acetate should be 1/10 of the total amount of the PCR products.

4. The H_2O control without template reveals possible false signals that arise from template independent primer extension due to self-annealing of primers. It will also reveal if the silicon rubber grid has been too loosely placed on the slide, which might result in leakage of the templates between wells.

5. The reaction wells should be kept humid during the incubation/ hybridization time since drying of the reaction mixture will result in high background fluorescence, which complicates the signal detection. A humid environment is accomplished for example by placing a wet tissue on the plexyglass lid.

6. The washings should be done avoiding exposure to light to avoid bleaching of the fluorophores.

7. In addition to the specific signal detection and analysis softwares that are available with the array scanner instruments the data interpretation becomes more effective using for example Microsoft® Excel-based macro.

8. An advantage of the "Tag-array" system is that generic micro-arrays can be manufactured in advance and used for many different applications. A second advantage of the approach is the increased sensitivity obtained through the cyclic minisequencing reactions. A disadvantage is that two oligonucleotide primers (the tagged minisequencing primer and the immobilized "anti-tag capture" primer) are needed for each SNP compared to only one in the minisequencing reaction using specific immobilized primers. The latter approach may also be easier to automate, and more advantageous for quantification of sequence variants.

References

1. Hacia, J. G. (1999) Resequencing and mutational analysis using oligonucleotide microarrays. *Nat. Genet.* **21**, 42–47.
2. Southern, E. M. (1996) DNA chips: analysing sequence by hybridization to oligonucleotides on a large scale. *Trends Genet.* **12**, 110–115.
3. Hacia, J. G., Sun, B., Hunt, N., Edgemon, K., Mosbrook, D., Robbins, C., et al. (1998) Strategies for mutational analysis of the large multiexon ATM gene using high-density oligonucleotide arrays. *Genome Res.* **8**, 1245–1258.
4. Gunderson, K. L., Huang, X. C., Morris, M. S., Lipshutz, R. J., Lockhart, D. J., and Chee, M. S. (1998) Mutation detection by ligation to complete n-mer DNA arrays. *Genome Res.* **8**, 1142–1153.
5. Pastinen, T., Kurg, A., Metspalu, A., Peltonen, L., and Syvänen, A. C. (1997) Minisequencing: a specific tool for DNA analysis and diagnostics on oligonucleotide arrays. *Genome Res.* **7**, 606–614.
6. Wang, D. G., Fan, J. B., Siao, C. J., Berno, A., Young, P., Sapolsky, R., et al. (1998) Large-scale identification, mapping, and genotyping

of single-nucleotide polymorphisms in the human genome. *Science* **280**, 1077–1082.

7. Cho, R. J., Mindrinos, M., Richards, D. R., Sapolsky, R. J., Anderson, M., Drenkard, E., et al. (1999) Genome-wide mapping with biallelic markers in Arabidopsis thaliana. *Nat. Genet.* **23**, 203–207.

8. Fortina, P., Delgrosso, K., Sakazume, T., Santacroce, R., Moutereau, S., Su, H. J., et al. (2000) Simple two-color array-based approach for mutation detection [In Process Citation]. *Eur. J. Hum. Genet.* **8**, 884–894.

9. Southern, E., Mir, K., and Shchepinov, M. (1999) Molecular interactions on microarrays. *Nat. Genet.* **21**, 5–9.

10. Syvänen, A.-C., Aalto-Setälä, K., Harju, L., Kontula, K., and Söderlund, H. (1990) A primer-guided nucleotide incorporation assay in the genotyping of apolipoprotein E. *Genomics* **8**, 684–692.

11. Pastinen, T., Raitio, M., Lindroos, K., Tainola, P., Peltonen, L., and Syvanen, A. C. (2000) A system for specific, high-throughput genotyping by allele-specific primer extension on microarrays. *Genome Res.* **10**, 1031–1042.

12. Raitio, M., Lindroos, K., Laukkanen, M., Pastinen, T., Sistonen, P., Sajantila, A., and Syvanen, A. C. (2001) Y-chromosomal Snps in finno-ugric-speaking populations analyzed by minisequencing on microarrays. *Genome Res.* **11**, 471–482.

13. Shoemaker, D. D., Lashkari, D. A., Morris D., Mittmann, M., and Davis, R. W. (1996) Quantitative phenotypic analysis of yeast deletion mutants using a highly parallel molecular bar-coding strategy. *Nat. Genet.* **14**, 450–456.

14. Cai, H., White, P. S., Torney, D., Deshpande, A., Wang, Z., Marrone, B., and Nolan, J. P. (2000) Flow cytometry-based minisequencing: a new platform for high-throughput single-nucleotide polymorphism scoring. *Genomics* **66**, 135–143.

15. Fan, J. B., Chen, X., Halushka, M. K., Berno, A., Huang, X., Ryder, T., et al. (2000) Parallel genotyping of human SNPs using generic high-density oligonucleotide tag arrays. *Genome Res.* **10**, 853–860.

16. Hirschhorn, J. N., Sklar, P., Lindblad-Toh, K., Lim, Y. M., Ruiz-Gutierrez, M., Bolk, S., et al. (2000) SBE-TAGS: An array-based method for efficient single-nucleotide polymorphism genotyping [In Process Citation]. *Proc. Natl. Acad. Sci. USA* **97**, 12,164–12,169.

17. Don, R. H., Cox, P. T., Wainwright, B. J., Baker, K., and Mattick, J. S. (1991) "Touchdown" PCR to circumvent spurious priming during gene amplification. *Nucleic Acids Res.* **19**, 4008.

18. Shuber, A. P., Grondin, V. J., and Klinger, K. W. (1995) A simplified procedure for developing multiplex PCRs. *Genome Res.* **5**, 488–493.
19. Zangenberg, G., Saiki, R.K., and Reynolds, R. (1999) Multiplex PCR: Optimization guidelines, in *PCR Applications* (Innis, M. A., Gelfand, D. H., Sninsky, J. J, ed.), Academic Press, London, UK, pp. 73–94.
20. Guo, Z., Guilfoyle, R. A., Thiel, A. J., Wang, R., and Smith, L. M. (1994) Direct fluorescence analysis of genetic polymorphisms by hybridization with oligonucleotide arrays on glass supports. *Nucleic Acids Res.* **22**, 5456–5465.

11

Quantitative Analysis of SNPs in Pooled DNA Samples by Solid-Phase Minisequencing

Charlotta Olsson, Ulrika Liljedahl,
and Ann-Christine Syvänen

1. Introduction

Quantitative analysis of pooled DNA samples is today generally recognized as a promising approach to determine allele frequencies of single nucleotide polymorphism (SNP) markers. The use of pooling will increase the genotyping throughput in population genetic studies or association studies, because very large sets of SNPs may be analyzed in a large number of individuals. The concept of DNA sample pooling to reduce labor and cost in an association study was first suggested by Arnheim et al. already in 1985. In that study Southern blot hybridization was used to quantify restriction fragment length polymorphisms (RFLP) in the human leukocyte antigen (HLA) class II locus *(1)*. In later studies allele frequencies of microsatellite markers were determined by analyzing pooled DNA samples using polymerase chain reaction (PCR), followed by size separation and quantification of the alleles *(2–4)*. We have found that the solid-phase minisequencing method is an ideal tool to determine the relative amounts of two closely related sequences that

From: *Methods in Molecular Biology, vol. 212:*
Single Nucleotide Polymorphisms: Methods and Protocols
Edited by: P-Y. Kwok © Humana Press Inc., Totowa, NJ

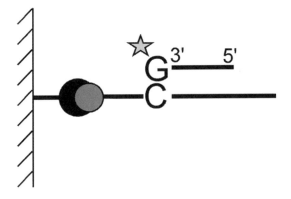

Fig. 1. Principle of the solid-phase minisequencing reaction. The minisequencing primer hybridizes to the immobilized single-stranded template, 3' adjacent to the variant nucleotide. The DNA polymerase extends the primer with the [³H]-labeled dNTP, if it is complementary to the nucleotide present at the variable site.

are present as a mixture in a DNA sample, such as the two alleles of a SNP in a pooled DNA sample *(5)*.

In the solid-phase minisequencing method a DNA fragment containing the site of the SNP or mutation is first amplified using one biotinylated and one unbiotinylated PCR primer. The biotinylated PCR products are captured on a streptavidin-coated solid support and denatured. The nucleotides at the SNP site are identified in the immobilized DNA by primer extension reactions, in which a DNA polymerase (*see* **Fig. 1**) incorporates a single labeled dNTP. Our first generation assay utilizes [³H]dNTPs as labels and microtiter plate wells as the solid support *(5)*. The results of the assay are numeric cpm-values expressing the amount of [³H]dNTP incorporated in the minisequencing reactions. The ratio between the cpm-values obtained in a minisequencing reaction (R-value) directly reflects the ratio between the two sequences in the original sample (*see* **Fig. 2**). When pooled DNA samples are analyzed, this ratio corresponds to the frequency of the SNP alleles among the individuals represented in the pooled sample. Because of the high sequence specificity of the

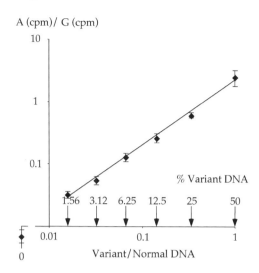

Fig. 2. Solid-phase minisequencing standard curve. The minisequencing signal ratio is plotted as a function of the ratio between the variant and normal sequences in the mixed sample. Mean value and standard deviations of four parallel assays are shown. The signal ratio obtained when analyzing a control sample with only normal sequence is shown on the axis on the left. Figure modified, from Olsson et al. *(7)*.

DNA polymerase-catalyzed incorporation of [^{3}H]-labeled dNTPs that are essentially identical to the natural dNTPs, the method allows quantitative determination of an allele present in the pooled sample at a frequency of less than 1% *(6–8)*. We have used minisequencing to accurately determine the allele frequencies of SNPs that are common in the population in large pooled samples containing DNA from hundreds or thousands of individuals *(5,7)*. **Table 1** shows the results from determining the allele frequencies of five polymorphisms in the ATP7B gene in a pooled DNA sample containing equal amounts of DNA from 2500 individuals *(7)*. The quantitative analysis was accurate with coefficient of variance (CV)-values lower than 10%, and the analysis of pooled samples gave a result concordant with the analysis of 20 individual samples. The protocol given below is the same as the protocol used in these analyses.

Table 1
Allele Frequencies of SNPs in the ATP7B Gene in a DNA Pool from 2500 Individuals and in 20 Random Individuals

SNP Sample	Signals (cpm)[a]		Signal ratio normal/variant	Allele distribution normal-variant	
	normal	variant		superpool	individual samples
G1216T					
Super pool	1120 ± 330	1940 ± 6	0.58	0.44–0.56	0.46–0.54
Heterozygote	1520 ± 100	2020 ± 150	0.75		
C1366G					
Super pool	1800 ± 170	1180 ± 210	1.52	0.52–0.48	0.50–0.50
Heterozygote	1870 ± 4	1340 ± 30	1.40		
G2855A					
Super pool	3500 ± 220	9460 ± 250	0.37	0.47–0.53	0.33–0.66
Heterozygote	3760 ± 250	9150 ± 560	0.41		
G2973A					
Super pool	840 ± 60	170 ± 30	4.94	0.92–0.08 / 0.94–0.06[b]	0.90–0.10
Heterozygote	820 ± 60	1820 ± 70	0.45		
5 % A-allele	1390 ± 180	250 ± 20	5.56		
G3045A					
Super pool	2000 ± 160	80 ± 8	25	0.96–0.04 / 0.97–0.03[b]	0.90–0.10
Heterozygote	1630 ± 130	1600 ± 180	1.01		
5% A-allele	1970 ± 310	150 ± 10	13.1		

The nucleotides denoted as "normal" are those given in the published sequence of the ATP7B gene (accession number U03464, *GenBank*). The normal nucleotide is given on the left, followed by the nucleotide number in the ATP7B sequence.
[a]Mean value and standard deviations of four parallel assays from the same PCR product.
[b]Calculated using a 5% allele mixture as reference. Data from Olsson et al. (7).

2. Materials

2.1. Equipment

1. Spectrophotometer with UV-light for measurement of DNA concentrations (*see* **Note 1**).
2. Thermocycler, and facilities to avoid contamination in PCR.
3. Microtiter plates with streptavidin-coated wells (e.g., Combiplate 8, Labsystems, Finland) (*see* **Note 2**).
4. Multichannel pipet and microtiter plate washer (optional).
5. Shaker at 37°C.
6. Water bath or incubator at 50°C.
7. Liquid scintillation counter.

2.2. Reagents

1. PCR primers designed according to standard procedures. One primer of each pair is biotinylated at its 5'-end.
2. Thermostable DNA polymerase (*see* **Note 3**).
3. dNTP mixture: 2 mM dATP, 2 mM dCTP, 2 mM dGTP, and 2 mM dTTP:
4. Phosphate-buffered saline (PBS)/Tween (capturing solution): 20 mM sodium phosphate buffer, pH 7.5, and 0.1% (v/v) Tween 20. 50 mL is sufficient for several full plate analyses.
5. TENT (washing solution): 40 mM Tris-HCl, pH 8.8, 1 mM EDTA, 50 mM NaCl, and 0.1% (v/v) Tween 20. 1–2 L is enough for several full-plate analyses.
6. 50 mM NaOH (make fresh every 4 wk).
7. Minisequencing primer. The primer is a 20-mer complementary to the biotinylated strand of the PCR product and designed to hybridize with its 3' end immediately adjacent to the variant nucleotide to be detected (*see* **Fig. 1**). The primer should be at least five nucleotides nested in relation to the unbiotinylated PCR primer.
8. [^{3}H]-labeled deoxynucleotides (dNTPs): dATP to detect a T at the variant site, dCTP to detect a G, etc. (Amersham; [^{3}H]dATP, TRK 625; dCTP, TRK 576; dGTP, TRK 627; dTTP, TRK 633), store at −20°C (*see* **Note 4**).
9. Scintillation reagent (for example Hi-Safe II, Wallac).

3. Method

3.1. Preparation of Pooled DNA Samples

DNA from blood samples is extracted using a standard procedure. The DNA concentration of the samples may be determined by measuring their absorbance at 260 nm in a spectrophotometer. Equal amounts of DNA from each individual are combined into a pooled sample (*see* **Note 1**).

3.2. PCR for Solid-Phase Minisequencing Analysis

Perform PCR according to routine protocols, except that the amount of the biotin-labeled primer should be reduced not to exceed the biotin-binding capacity of the microtiter well (*see* **Note 2**). For a 50 µL PCR reaction we use 10 pmol of biotin-labeled primer and 50 pmol of the unbiotinylated primer. To be able to use [^{3}H]dNTPs, which have low specific activities, for the minisequencing analysis, 1/10 of the PCR product should be visible after agarose gel electrophoresis and staining with ethidium bromide.

3.3. Solid-Phase Minisequencing Analysis

1. Affinity capture: Transfer 10 µL aliquots of the PCR product and 40 µL of PBS/Tween to two streptavidin-coated microtiter wells (*see* **Note 5**). Include as negative controls two wells without PCR product. Seal the wells with a sticker and incubate the plate at 37°C for 1.5 h with gentle shaking.
2. Wash the wells six times at room temperature by adding 200 µL of TENT to each well, discard the washing solution and empty the wells thoroughly between the washing steps (*see* **Note 6**).
3. Denature the captured PCR product by adding 100 µL of 50 m*M* NaOH to each well, followed by incubation at room temperature for 3 min. Discard the NaOH and wash the wells as in **step 2**.
4. For each DNA fragment to be analyzed, prepare two 50 µL mixtures of nucleotide-specific minisequencing solution, one for detection of the normal and one for the variant nucleotide, by mixing 5 µL of 10X DNA polymerase buffer, 10 pmol of detection step primer (for

example 2 µL of 5 µM primer), 0.1 µCi (usually 0.1 µL) of one [^{3}H]dNTP, 0.1 U of DNA polymerase, and H$_2$O to a total volume of 50 µL. It is obviously convenient to prepare master mixes for the desired number of analyses with each nucleotide (*see* **Note 7**).

5. Add 50 µL of one nucleotide-specific mixture to each well, and incubate the plate at 50°C for 10 min (*see* **Note 8**).
6. Discard the content of the wells and wash them as in **step 2**.
7. Release the detection step primer from the template by adding 60 µL 50 mM NaOH and incubate for 3 min at room temperature.
8. Measure the amount of [^{3}H]dNTP by which the primers have been extended in a liquid scintillation counter (*see* **Note 9**).
9. The result of the assay is obtained as cpm-values. The cpm-value of each reaction expresses the amount of the incorporated [^{3}H]dNTP. Calculate the ratio (R) (*see* **Table 1** and **Note 10**):

$$R = \frac{cpm\ incorporated\ in\ the\ reaction\ for\ variant\ nucleotide}{cpm\ incorporated\ in\ the\ reaction\ for\ normal\ nucleotide}$$

3.4. Preparation of a Standard Curve

Mix DNA from individuals of known genotypes into different proportions of the normal and variant alleles ranging from 0.5– 99.5% of variant sequence, and analyze the mixtures in parallel with the pooled DNA samples. Plot the R-values obtained in the minisequencing reaction on a log-log scale as a function of the ratio between the sequences present in the original mixture to obtain a linear standard curve (*see* **Fig. 2**). The standard curve is then utilized to accurately determine the ratio between the two alleles in a pooled DNA sample. Alternatively, a sample from a heterozygous individual may be used as a reference sample for quantification. The allele frequencies (f) of the SNPs may be calculated by comparing the signal ratios observed in the large pooled sample (R$_{pool}$) to the corresponding signal ratios in heterozygous samples (R$_{Het}$), where the two alleles are present at a 1:1 ratio (50%) according to the formulas:

$$f_{allele1} = \frac{R_{pool}/R_{Het}}{1 + R_{pool}/R_{Het}} \quad ; \quad f_{allele2} = \frac{1}{1 + R_{pool}/R_{Het}}$$

4. Notes

1. It is crucial for accurate determination of allele frequencies in pooled DNA samples that the pool contains an equal amount of DNA from each individual. Inclusion of as many DNA samples as possible in the pools reduces the error introduced by measurement of the DNA concentration of the samples to be pooled.

2. The binding capacity of the streptavidin-coated microtiter well (Labsystems) is 2–5 pmol of biotinylated oligonucleotide. Therefore a reduced amount of biotinylated primer is used for PCR. If a higher binding capacity is desired, for example streptavidin-coated magnetic polystyrene beads (Dynabeads M-280, Dynal, Norway) can be used.

3. The use of a thermostable DNA polymerase in the minisequencing primer extension reaction is advantageous, since a high temperature, favorable for the simultaneous primer annealing reaction, can be used.

4. The [^{3}H]dNTPs are weak β-emitters and their half lives are long (13 yr). The necessary precautions for working with [^{3}H] should be taken. Also dNTPs or dideoxy-nucleotides labeled with fluorophores *(9)* or colorimetrically detectable haptens *(10)* can be used at the cost of sensitivity of detecting minority alleles in the pooled samples.

5. Each nucleotide to be detected at the variant site is analyzed in a separate well. Thus at least two wells are needed per PCR product. For quantitative applications we carry out two (or more) parallel assays for each nucleotide, i.e., four wells per PCR product.

6. Washing can be performed in an automated microtiter plate washer, or by manually pipeting the washing solution to the wells, discarding the liquid and tapping the plate against a tissue paper. It is important for the specificity of the minisequencing reaction to thoroughly empty the wells between the washing steps, to remove completely all dNTPs from the PCR. The presence of other dNTPs than the intended [^{3}H]dNTP during the minisequencing reaction will cause unspecific extension of the detection step primer.

7. The minisequencing reaction mixture without the polymerase, can be kept at room temperature for 1–2 h. It is convenient to prepare it during the incubation in **step 1**.

8. The conditions for hybridizing the minisequencing primer are not stringent, and the temperature of 50°C can be applied to analysis of most SNPs irrespective of their sequence context.

9. Streptavidin-coated microtiter plates made of scintillating polystyrene are available (ScintiStrips, Wallac, Finland). When these plates

are used, the final washing, denaturation, and transfer of the eluted detection primer to scintillation vials can be omitted, but a scintillation counter for microtiter plates is needed *(11)*.

10. The ratio between the cpm-values for the two nucleotides reflects the ratio between the two sequences in the original sample. The R-value is affected by the specific activities of the [^{3}H]dNTPs used, and if more than one identical [^{3}H]dNTP will be incorporated in the sequence following the SNP site in either the normal or the variant allele this will obviously also affect the R-value. Both of these factors can easily be corrected, when calculating the ratio between the two sequences. Alternatively, a standard curve or a heterozygous reference sample can be used to correct for these factors (*see* **Subheading 3.4.**).

References

1. Arnheim, N., Strange, C., and Erlich, H. (1985) Use of pooled DNA samples to detect linkage disequilibrium of polymorphic restriction fragments and human disease: studies of the HLA class II loci. *Proc. Natl. Acad. Sci USA* **82**, 6970–6974.

2. Lipkin, E., Mosig, M. O., Darvasi, A., Ezra, E., Shalom, A., Friedmann, A., and Soller, M. (1998) Quantitative trait locus mapping in dairy cattle by means of selective milk DNA pooling using dinucleotide microsatellite markers: analysis of milk protein percentage. *Genetics* **149**, 1557–1567.

3. Pacek, P., Sajantila, A., and Syvänen, A. C. (1993) Determination of allele frequencies at loci with length polymorphism by quantitative analysis of DNA amplified from pooled samples. *PCR Methods Appl.* **2**, 313–317.

4. Shaw, S. H., Carrasquillo, M. M., Kashuk, C., Puffenberger, E. G., and Chakravarti, A. (1998) Allele frequency distributions in pooled DNA samples: applications to mapping complex disease genes. *Genome Res.* **8**, 111–123.

5. Syvänen, A.-C., Sajantila, A., and Lukka, M. (1993) Identification of individuals by analysis of biallelic DNA markers, using PCR and solid-phase minisequencing. *Am. J. Hum. Genet.* **52**, 46–59.

6. Syvänen, A.-C., Ikonen, E., Manninen, T., Bengtstrom, M., Soderlund, H., Aula, P., and Peltonen, L. (1992) Convenient and quantitative determination of the frequency of a mutant allele using solid-phase minisequencing: application to aspartylglucosaminuria in Finland. *Genomics* **12**, 590–595.

7. Olsson, C., Waldenström, E., Westermark, K., Landegren, U., and Syvänen, A. C. (2000) Determination of the frequencies of ten allelic variants of the Wilson disease gene (ATP7B), in pooled DNA samples. *Eur. J. Hum. Genet.* **8**, 933–938.

8. Lagerström, M., Olsson, C., Forsgren, L., and Syvänen, A.-C. (2001) Heteroplasmy of the human mitochondrial DNA control region remains constant during life. *Am. J. Hum. Genet.* **68**, 1299–1301.

9. Pastinen, T., Partanen, J., and Syvänen, A.-C. (1996) Multiplex, fluorescent, solid-phase minisequencing for efficient screening of DNA sequence variation. *Clin Chem.* **42**, 1391–1397.

10. Nikiforov, T. T., Rendle, R. B., Goelet, P., Rogers, Y. H., Kotewicz, M. L., Anderson, S., et al. (1994) Genetic Bit Analysis: a solid phase method for typing single nucleotide polymorphisms. *Nucleic Acids Res.* **22**, 4167–4175.

11. Ihalainen, J., Siitari, H., Laine, S., Syvänen, A.-C., and Palotie, A. (1994) Towards automatic detection of point mutations: use of scintillating microplates in solid-phase minisequencing. *Biotechniques* **16**, 938–943.

12

Homogeneous Primer Extension Assay With Fluorescence Polarization Detection

Tony M. Hsu and Pui-Yan Kwok

1. Introduction

The primer extension assay with fluorescence polarization (FP) is a genotyping method that combines the specificity of nucleotide incorporation by DNA polymerase and the sensitivity of fluorescence polarization. We named the assay Template-directed Dye-terminator Incorporation assay with FP detection (FP-TDI assay). It is a dideoxy chain-terminating DNA-sequencing protocol that ascertains the nature of the one base immediately 3' to the sequencing primer (also called single nucleotide polymorphism [SNP]-specific primer). The SNP-specific primer is designed to anneal immediately upstream of the polymorphic site on the target DNA. In the presence of the target DNA, the appropriate dye-labeled terminators, and DNA polymerase, the SNP-specific primer is extended by one base as dictated by the nature of the allele at the polymorphic site on the target DNA. By determining which terminator is incorporated, the allele present in the target DNA can be inferred (**Fig. 1**). Template-directed primer extension reaction has been used in various formats for genotyping, and it has proved to be highly specific and sensitive *(1–4)*.

From: *Methods in Molecular Biology, vol. 212:*
Single Nucleotide Polymorphisms: Methods and Protocols
Edited by: P-Y. Kwok © Humana Press Inc., Totowa, NJ

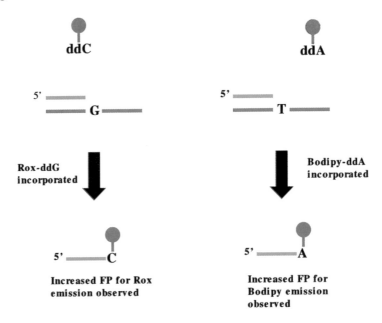

Fig. 1. Primer extension reaction with fluorescence polarization detection. With the SNP probe annealing to the target DNA next to the polymorphic site, DNA polymerase incorporates the specific dideoxy- (or acyclo-) nucleoside triphosphate labeled with a fluorescent dye onto the probe to yield a substantially larger fluorescent molecule, which has a much higher fluorescence polarization value than that of the free nucleoside triphosphate. In the panel *at left*, ddC is incorporated onto the SNP probe hybridized to the PCR product with the "G" allele. In the panel *at right*, ddA is incorporated onto the probe hybridized to the PCR products with the "T" allele.

FP is based on the observation that when a fluorescent molecule is excited by plane-polarized light, the fluorescent light it emits is also polarized *(5)*. However, because the fluorescent molecule rotates and tumbles in solution at room temperature, FP is not fully observed when monitored by an external detector. The FP of a molecule is proportional to its rotational relaxation time (the time it takes to rotated through 68.5°). This value is related to the viscosity of the solvent, absolute temperature, and molecular volume. If the viscosity and absolute temperature are held constant, FP is proportional to the molecular volume, which is directly proportional to the molecular weight. Therefore, FP is an ideal detection format for methods, such as the TDI assay, that involve significant changes in the molecular weight of the fluorescent molecule.

FP is expressed in the ratio of fluorescence detected in the vertical and horizontal axes, and is therefore independent of fluorescence intensity. This is a clear advantage over other methods that rely on fluorescence intensity in that with FP, sample-to-sample variation in the amount of fluorescent dye does not affect the results and so no reference dye is needed. The total polarization reflects the sum of FP from all fluorescent species in solution as described by the equation: $P = P_{max}[ddNTP]_b + P_{min}([ddNTP]_i-[ddNTP]_b)$ where P_{max} is the polarization for dye-labeled ddNTP incorporated onto the SNP-specific primer, P_{min} is the polarization of the unincorporated dye-labeled dd-NTP, $[ddNTP]_i$ is the initial concentration of dye-labeled ddNTP, and $[ddNTP]_b$ is the concentration of incorporated dye-labeled ddNTP. The maximum change in FP is achieved when all the ddNTP are incorporated unto the TDI primer. Therefore, it is important that the initial concentration of dye-labeled ddNTP used in the reaction is kept at a minimum.

FP-TDI assay utilizes four spectrally distinct dye-terminators to achieve universal assay conditions. Even without optimization, about 80% of all SNP markers tested yielded robust results. Addition of *Escherichia coli* single-stranded DNA binding protein (SSB) just prior to the FP reading increased FP values of the products significantly and brought the success rate of FP-TDI assays up to 90%. With further modifications of the assay we were able to improve the assay to achieve a 100% success rate (*6–8*).

2. Materials

2.1. PCR

1. Thermostable DNA polymerase AmpliTaq Gold™ at 5 U/µL (Applied Biosystems, Foster City, CA).

2. GeneAmp 10X PCR Buffer II: 100 mM Tris-HCl, pH 8.3, 500 mM KCl (Applied Biosystems).
3. 25 mM MgCl$_2$ Solution (Applied Biosystems).
4. dNTP mixture: 2.5 mM dATP, 2.5 mM dCTP, 2.5 mM dGTP, and 2.5 mM dTTP.
5. PCR primers are designed according to standard procedures by Life Technologies (Gaithersburg, MD).

2.2. Degradation of Excess PCR Primer and dNTPs

1. Shrimp alkaline phosphatase (SAP) 1000 U (Roche, Mannheim, Germany, Cat. no. 1 758 250).
2. Exonuclease I 2500 U (USB Corporation, Cleveland, OH, Cat. no. 70073Z).
3. Dephosphorylation buffer: 0.5 M Tris-HCl, 50 mM MgCl$_2$, pH 8.5 (supplied by Roche when you buy shrimp alkaline phosphatase).

2.3. Primer Extension

1. Thermo Sequenase DNA Polymerase 1000 U. Dilute the polymerase from 32 U/µL to 8 U/µL prior to use (Amersham Pharmacia Biotech, Piscataway, NJ, Cat. No. E79000Y).
2. Rox-ddGTP, Bodipy-Fluorescein-ddATP, Tamra-ddCTP and R6G-ddUTP (Perkin Elmer, Inc., Boston, MA, Cat. no. NEL479 for Rox-ddGTP, NEL574 for Bodipy-Fluorescein-ddATP, NEL473 for Tamra-ddCTP, and NEL488 for R6G-ddUTP).
3. TDI buffer: 250 mM Tris-HCl pH 9.0, 250 mM KCl, 25 mM NaCl, 25 mM MgCl$_2$, 40% glycerol.
4. SNP-specific primer: this primer is a 20- to 25-mer complementary to either the sense or antisense strand of the target DNA and designed to anneal with its 3'-end immediately adjacent to the polymorphic site. It is synthesized at PCR primer grade by Invitrogen (Gaithersburg, MD).
5. [Optional] Single-Stranded DNA Binding Protein (SSB) 500 µg (USB Corporation, Cleveland OH, Cat. no. 70032Z).

2.5. Equipment

1. 96-Well black PCR plates. All reactions are run and read in these plates (Marsh Bioproducts, Rochester, NY, Cat. no. AB-0800/black).
2. Sealing mats (Fisher Scientific, Cat. no. 07-200-614).

3. Thermocyclers.
4. Victor2 plate reader (PerkinElmer) for FP reading (*see* **Note 1**).

3. Methods

3.1. PCR Amplification

1. Assemble 10 μL reaction mixtures containing 5 μL of genomic DNA (a total of 5 ng), 1 μL of GeneAmp 10X PCR Buffer II, 1 μL of 25 mM MgCl$_2$ solution, 0.2 μL of 2.5 mM dNTP mixture, 0.5 μL of 2.5 μM each of PCR primers, 0.05 μL (0.25 U) of AmpliTaq Gold DNA polymerase, and 1.75 μL of ddH$_2$O (*see* **Notes 2–5**).
2. Heat the reaction mixture at 95°C for 10 min to activate the AmpliTaq Gold DNA polymerase followed by 35 amplification cycles. Each cycle consists of denaturation at 93°C for 30 s, primer annealing at 58°C for 40 s, and primer extension at 72°C for 40 s.
3. Incubate the reaction mixtures at 72°C for 5 min for final primer extension.
4. At the end of the reaction, hold the product mixture at 4°C until further use.

3.2. Degradation of Excess PCR Primers and dNTPs

1. At the end of the PCR assay, add 5 μL of an enzymatic cocktail to the PCR mixture. This enzymatic cocktail contains 0.2 μL (0.2 U) of shrimp alkaline phosphatase, 0.1 μL (1 U) of *Escherichia coli* exonuclease I, 1 μL of dephosphorylation buffer, and 3.7 μL of ddH$_2$O.
2. Incubate the mixture at 37°C for 45 min.
3. Heat inactivate the enzymes at 95°C for 15 min.
4. Keep the reaction mixture at 4°C until further use.

3.3. Single Base Extension

1. Add 10 μL of TDI cocktail to the enzymatically treated PCR product. The TDI cocktail contains 2 μL of TDI buffer, 1 μL of 10 μM SNP-specific primer, 0.05 μL (0.4 U) of Thermo Sequenase DNA polymerase, 0.05 μL of 4-dye-ddNTP mixture (mix equal volume of 25 μM Rox-ddGTP, 25 μM Bodipy-Fluorescein-ddATP, 25 μM Tamra-ddCTP, and 25 μM R6G-ddUTP to make the 4-dye-ddNTP mixture), and 6.9 μL of ddH$_2$O (*see* **Notes 3** and **6**).

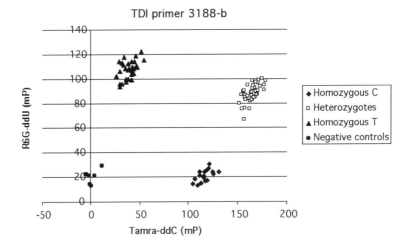

Fig. 2. Results of a TDI assay typed on 96 samples. This is a typical assay in which the FP values of the samples group into four clusters. The cluster near the origin, where FP values are low for both dyes, consists of negative (no DNA) controls and failed samples (no PCR product formed). The cluster in the lower right corner, where FP values for TAMRA-C are high but for R6G-U are low, consists of samples that are homozygous for the "C" allele. The cluster in the upper left corner, where FP values for TAMRA-C are low but for R6G-U are high, consists of samples that are homozygous for the "T" allele. The cluster in the upper right corner, where FP values are high for both dyes, consists of heterozygous samples.

2. Incubate the reaction mixture at 93°C for 1 min, followed by 50 cycles of 93°C for 10 s and 50°C for 30 s.
3. At the end of the reaction, hold the samples at 4°C.

3.4. Data Analysis

1. Export the FP values for two dyes to a spread sheet (such as Microsoft Excel) and use it to create an X-Y scatter plot.
2. Assign the genotype of each sample by noting its position in the plot. The no DNA negative controls have low FP values for both dyes analyzed and occupy the area near the origin of the plot, indicating that the small dye-terminators remain free in solution (*see* **Fig. 2** as an example). For homozygous individuals, the FP values for one of the dyes is high while that for the other dye is low. In the example, the

TAMRA-ddC FP values for the homozygotes for the C allele are high and the values for R6G-ddU are low, reflecting significant incorporation of the TAMRA-ddC terminator onto the SNP-specific primer but minimal incorporation of the R6G-ddU. Conversely, the FP values of TAMRA-ddC are low and those of R6G-ddU are high for homozygous T individuals. As for heterozygous individuals, the FP values for both TAMRA-ddC and R6G-ddU are high because of significant incorporation of both dye-terminators (*see* **Notes 7** and **8**).

3.5. E. coli *Single-Stranded DNA Binding Protein (SSB) (Optional)*

1. After the primer extension reaction, add 10 μL of a mixture containing 0.1 μL of 2.2 μg/μL SSB, 2 μL of TDI buffer, and 7.9 μL of ddH$_2$O to the mixture.
2. Incubate at 37°C for 1 h and hold at 4°C until use.
3. After SSB incubation, add 75 μL of reading buffer (consisting of 25 μL 95% ethanol, 10 μL of TDI buffer, and 40 μL of ddH$_2$O) to each reaction before reading the plates.

4. Notes

1. We have also used the LJL Analyst plate reader (Molecular Devices, Sunnyvale, CA). Other laboratories have tried and recommended other plate readers, such as Tecan Ultra from Tecan (Männedorf, Switzerland).
2. For the PCR amplification step, do not use more dNTP and PCR primers than indicated. Excess dNTP and PCR primers increase nonspecific incorporation in the single base extension step.
3. PCR primers and SNP-specific primers are designed using PrimerExpress. However, other programs such as Primer 3 can be used for this purpose. PCR products ranging in size from 80–600 bp have worked for FP-TDI assay, but we usually use PCR products between 80 bp and 250 bp to ensure maximum PCR efficiency. When designing SNP-specific primers, we first make sure that the T$_m$ of the primers is greater than 60°C, and then we try to use SNP-specific primers with a length of greater than 24 bp. Both sense and antisense SNP-specific primers are usually ordered for each marker so that both primers can be tested to see which one works best.

4. To avoid problems with cross-contamination, we add the PCR mixture into the wells first before we add the genomic DNA samples.

5. To minimize contamination from transfer, we use Marsh 96-well black PCR plate through the entire FP-TDI assay including FP reading in the LJL plate reader. Other PCR plates or PCR tubes can be used for all the steps before FP reading. However, black plates have to be used for FP reading in LJL plate reader.

6. The AcycloPrime-FP SNP Detection Kit from PerkinElmer's NEN Division can also be used for the single base extension step. In this kit, acycloterminators instead of dideoxynucleotide terminators are used in the single base extension step. AcycloPol, a novel mutant thermostable polymerase from the Archeon family, is used in this new kit because it has a higher affinity and specificity for acycloterminators than various Taq mutants have for dideoxynucleotide terminators. Extensive data show that AcycloPrimer-FP SNP detection Kit is at least as robust as the current FP-TDI protocol in SNP genotyping.

7. For about 80% of the SNP markers, either the sense or antisense or both SNP-specific primers would give clear-cut genotype data. However, in rare instances, neither SNP-specific primer works. In these cases, the SSB step and reading buffer step are added to improve the results. **Figure 3A** shows the results before the addition of SSB and reading buffer. The addition of SSB and reading buffer greatly improves the data (*see* **Fig. 3B**).

8. Even with SSB and reading buffer, some markers would still give unsatisfactory results. All of these failures are secondary to misincorporation of dye-terminator as shown in **Fig. 4A** where the cluster for homozygous A and the cluster for heterozygotes merge. This happens because the dye-terminator for the correct allele (in this case, Rox-ddG) was used up prematurely and the wrong dye-terminator (Bodipy-ddA) was incorporated instead. To correct this problem, two modifications can be done: 1) Add more dye-terminators (add 0.1 µL of dye-terminator mix for each reaction instead of 0.05 µL); 2) Reduce the number of TDI cycles (from 50 cycles to 25 cycles). Or if needed, both modifications can be performed at the same time to improve the results. **Figure 4B** shows a significant improvement of data after reducing the number of TDI cycles from 50 to 25 and doubling the dye-terminators. As mentioned in **Note 7**, most TDI primers work well with our standard protocol stated in the

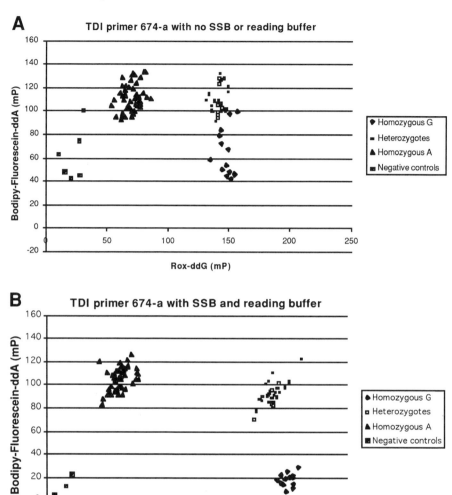

Fig. 3. Effect of SSB protein on TDI assay results. For a small number of SNPs, secondary and/or tertiary structures of the dye-labeled probes are such that the FP values are lower than usual, making the separation between incorporated and free dyes too small for confident genotype calls. In **(A),** the clusters are easily distinguished for the TAMRA dye but poorly separated for the R6G dye. By adding SSB protein to the mixture, the probes coated with SSB protein are straightened out and heavier than before, yielding products that are now well-separated from the free dyes **(B)**.

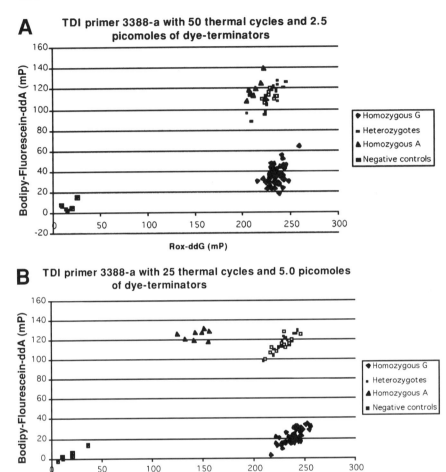

Fig. 4. Effect of reduced number of primer extension cycles on TDI assay results. On rare occasions, extremely high yields of PCR lead to the exhausting of one dye terminator and mis-incorporation of the second dye terminator. In (**A**), the bodipy-A terminators are used up early in the reaction and ROX-G terminators are incorporated onto the SNP probes even when they are hybridized to the PCR products containing the "A" allele. In this situation, only three clusters are observed, with the upper left corner devoid of samples because the homozygous "A" samples are now merged with the heterozygotes. By reducing the number of primer extension cycles and increasing the dye-terminator concentration (both measures ensure that the terminators are not exhausted during the reaction), the clusters are clearly separated (**B**).

Methods section. Only very few of them require modifications as described here. We do not yet know why some markers require modifications and others don't. We suspect that it has to do with the nucleotide sequence around the SNP markers; we will need more data to come to a better understanding of this phenomenon.

Acknowledgment

This work is supported by grants from the National Institutes of Health (RO1-EY12557 to PYK and T32-AR07284 to TMH). We thank S. Duan for technical assistance, PerkinElmer NEN for generous gifterous gifts of dye-terminators, and Applied Biosystems for generous gift of a DNA polymerase.

References

1. Chen, X., Levine, L., and Kwok, P.-Y. (1999) Fluorescence polarization in homogeneous nucleic acid analysis. *Genome Res.* **9**, 492–498.
2. Nikiforov, T. T., Rendle, R. B., Goelet, P., Rogers, Y. H., Kotewicz, M. L., Anderson, S., et al. (1994) Genetic bit analysis: a solid phase method for typing single nucleotide polymorphisms. *Nucleic Acids Res.* **22**, 4167–4175.
3. Syvänen, A.-C. (1994) Detection of point mutations in human genes by the solid-phase minisequencing method. *Clin. Chim. Acta* **226**, 225–236.
4. Pastinen, T., Kurg, A., Metspalu, A., Peltonen, L., and Syvänen, A.-C. (1997) Minisequencing: a specific tool for DNA analysis and diagnostics on oligonucleotide arrays. *Genome Res.* **7**, 606–614.
5. Perrin, F. (1926) Polarization de la lumiere de fluorescence. Vie moyenne de molecules dans l'etat excite. *J. Phys. Radium* **7**, 390–401.
6. Hsu, T. M., Chen, X., Duan, S., Miller, R., and Kwok, P.-Y. (2001) Universal SNP genotyping assay with fluorescence polarization detection. *Biotechniques* **31**, 560, 562, 564–568.
7. Raghunathan, S., Kozlov, A. G., Lohman, T. M., and Waksman, G. (2000) Structure of the DNA binding domain of E. coli SSB bound to ssDNA. *Nature Struc. Biol.* **7(8)**, 648–652.
8. Chrysogelos, S. and Griffith, J. (1982) Escherichia coli single-strand binding protein organizes single-stranded DNA in nucleosome-like units. *Proc. Natl. Acad. Sci. USA* **79**, 5803–5807.

13

Pyrosequencing for SNP Genotyping

Mostafa Ronaghi

1. Introduction

Pyrosequencing is a new DNA sequencing technique based on sequencing-by-synthesis *(1)*. This technique enables real-time detection using an enzyme-cascade system, consisting of four enzymes and specific substrates, to produce light whenever a nucleotide forms a base pair with the complementary nucleotide in a DNA template strand. As a result of nucleotide incorporation inorganic pyrophosphate (PPi) is released and is subsequently converted to ATP by ATP sulfurylase which is used by luciferase to generate proportional amount of light. Unreacted nucleotides are degraded by the enzyme apyrase, allowing iterative addition of nucleotides (*see* **Fig. 1**). DNA template generated by PCR is hybridized with a sequencing primer prior to Pyrosequencing. Using one pmol of DNA, 6×10^{11} ATP molecules can be obtained per nucleotide incorporation which, in turn, generate more than 6×10^9 photons at a wavelength of 560 nanometers. This amount of light is easily detected by a photodiode, photomultiplier tube, or a CCD-camera. Pyrosequencing has the potential advantages of accuracy, flexibility, parallel processing, and simple automation. Furthermore, the technique avoids the use of labeled primers, labeled nucleotides,

From: *Methods in Molecular Biology, vol. 212:*
Single Nucleotide Polymorphisms: Methods and Protocols
Edited by: P.-Y. Kwok © Humana Press Inc., Totowa, NJ

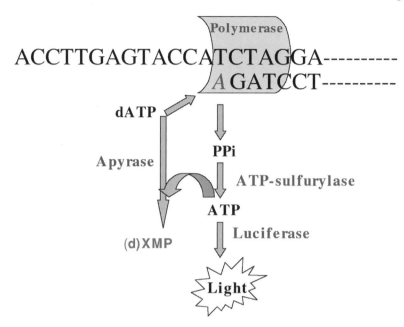

Fig. 1. Schematic representation of the progress of the enzymatic reaction in Pyrosequencing. DNA template with hybridized primer and four enzymes involved in Pyrosequencing are added to a well of microtiter plate. The four different nucleotides are added stepwise and incorporation is followed using the enzyme ATP sulfurylase and luciferase. The nucleotides are continuously degraded by apyrase allowing addition of subsequent nucleotide. (d)XMP indicates both AMP and dNMP.

and gel-electrophoresis. The methodological performance of this technique in sequence determination of difficult secondary DNA structure *(2)*, cDNA sequencing *(3)*, re-sequencing of disease genes *(4)*, microbial typing (Pourmand et al., unpublished results, Gharizadeh, unpublished results), and SNP genotyping *(5–7)* has been demonstrated and was recently improved by addition of single-stranded DNA binding protein *(8)*. This chapter details the steps involved in template preparation and Pyrosequencing and the use of this technique for SNP analyses.

2. Materials

2.1. Preparation of the Template

1. Templates for Pyrosequencing are usually generated by PCR in which one of the primers is biotinylated (*see* **Note 1**).
2. Streptavidin-coated magnetic beads.
3. Binding/washing buffer: 10 mM Tris-HCl, pH 8.0, 2 M NaCl, and 0.1% Tween 20. For immobilization of biotinylated PCR products onto magnetic beads.
4. Sequencing primer (one of the PCR primer can be used as sequencing primer if it is close to the SNP position.

2.2. Pyrosequencing Reaction

1. Enzyme mix: 5 U exonuclease-deficient Klenow DNA polymerase, 40 mU apyrase, 500 ng purified luciferase, 15 mU purified recombinant ATP sulfurylase. The enzyme mix can be lyophilized for long-term storage and diluted before use.
2. Substrate mix: 5 mM magnesium acetate, 0.1% bovine serum albumin (BSA), 1 mM dithiothreitol (DTT), 5 µM adenosine 5'-phosphosulfate (APS), 0.4 mg/mL polyvinylpyrrolidone (360,000), and 100 µg/mL D-luciferin. The substrate mix, which is light-sensitive, can be lyophilized or stored at −20°C for at least 1 yr.
3. Twenty microliters of 0.1 M nucleotides dATP-αS, dCTP, dGTP, and dTTP are each treated with 1 µL of 100 mM inorganic pyrophosphatase to remove the unspecific signals generated by contaminated PPi (*see* **Note 2**). The nucleotides are diluted to 1000 µL and can be stored at 4° to −20°C for at least 1 yr.
4. Microtiter plate: If PSQ 96 system for Pyrosequencing is used, special plates need to be used.
5. Inkjet cassette: for delivery of enzymes and substrates using PSQ 96 system.

All the described products are commercially available from Pyrosequencing AB.

2.3. Equipment

1. Pyrosequencing machine: two machines are now commercially available PSQ 96, and PTP 384 (Pyrosequencing AB).
2. Magnetic rack for sedimentation of the magnetic beads.

3. Methods

3.1. Preparation of the Template for Pyrosequencing

1. Immobilize the biotinylated PCR product (40–50 µL) onto 200 µg streptavidin-coated paramagnetic beads in 40–50 µL binding washing buffer for 15 min at 43°C with mild agitation.
2. Centrifuge the beads and remove the supernatant. Incubate the immobilized template in 20 µL 0.1 M NaOH for 5 min at room temperature or at 43°C to obtain pure single-stranded DNA.
3. After washing, resuspend the immobilized strand in the annealing buffer (10 mM Tris-HCl, 2 mM MgAC$_2$, pH 7.5) containing 10 pmol of sequencing primer in a total volume of 10 µL.
4. Hybridize the sequencing primer to the template by incubating at 94°C for 20 s, 65°C for 2 min and then cooled to room temperature (*see* **Note 3**).

3.2. Pyrosequencing Reaction

1. Transfer DNA template with hybridized primer to a microtiter plate from Pyrosequencing and diluted to 50 µL with TAE buffer (100 mM Tris-Acetate, pH 7.75, 0.5 mM EDTA).
2. Add enzyme mix, substrate mix and nucleotides to the inkjet cartridge and place in Pyrosequencing machine.
3. The Pyrosequencing system will automatically start the reaction by addition of enzyme mix, substrate mix, and the nucleotides according to the specified order appropriate for the SNP to be analyzed. The SNP will be scored automatically.

3.3. Pyrosequencing Machine

The automated Pyrosequencing machine uses a disposable inkjet cartridge for precise delivery of submicroliter volumes of six different reagents into a temperature controlled microtiter plate *(9)*. The microtiter plate is under continuous agitation to increase the rate of the reactions. A lens array is used to efficiently focus the generated

A homozygote   **A/G heterozygote**   **G homozygote**

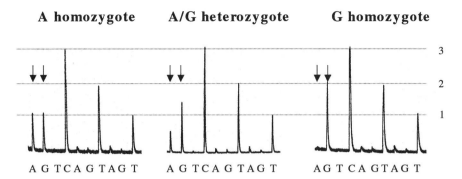

Fig. 2. Pyrogram of the raw data obtained from Pyrosequencing of three different genotypes. The order of nucleotide addition is indicated below the pyrogram. The arrows indicate the position of the polymorphisms. The numbers represent the height of the signals as a result of incorporation of one, two, or three nucleotides.

luminescence from each individual well of the microtiter plate onto the chip of a CCD-camera. Nucleotides are dispensed into alternating wells with a delay to minimize the cross-talk of generated light between adjacent wells. A cooled CCD-camera images the plate every second to follow the exact process of the Pyrosequencing reaction. Data acquisition modules and an interface for PC-connection are used in this instrument. The software running under Windows enables individual control of the nucleotide dispensation order for each well. Currently two automated versions of Pyrosequencing are available, i.e., PSQ 96 for simultaneous analysis of 96 samples and PTP 384 for simultaneous analysis of 384 samples. The latter machine can also be integrated with a Genesis workstation and a Te-Mags separation module enabling the analysis of up to 50,000 SNPs per day.

3.4. Data Analysis

Raw data can be obtained using the Pyrosequencing machine. Pyrosequencing data is quantitative because the amount of light generated is proportional to the amount of the PPi molecule which, in turn, is dependent only on number of incorporated nucleotide and independent of the nature of the nucleotides. Thereby the signals are proportional to the number of incorporated nucleotide(s) (*see* **Fig. 2**).

The proportionality of the data has provided the basis for development of different softwares including SNP software, Tag software, Allele quantification software, multiplex SNP genotyping software, and haplotyping software (under development) (*see* **Notes 4–6**).

4. Notes

1. The amount of the DNA template is crucial for successful Pyrosequencing. For Pyrosequencing using PSQ 96 system at least one pmol template is needed.
2. Purity of the nucleotides improves the correct calling of the sequences. Contaminants such as PPi should be removed before using in Pyrosequencing. It should be noted that in Pyrosequencing dATP-αS is used instead of dATP since it is silence for luciferase and is efficiently incorporated by DNA polymerase *(10)*.
3. Heat-sensitive enzymes (especially luciferase) in the Pyrosequencing reaction mixture require that the extension temperature be set to 28°C. Such conditions indisputably favor interactions within and between sequencing primer and template molecules. Oligonucleotides employed as sequencing primers must accordingly be carefully designed to avoid illegitimate extensions. Although the unspecific signals can be extensively eliminated by using single-stranded DNA binding protein *(8)*.
4. Pyrosequencing has the inherent problem of correct calling of homopolymeric regions due to the nonlinear light response following incorporation of more than 5–6 identical nucleotides.
5. Out of phase signals can be generated in *de novo* sequencing of heterozygous DNA regions. Use of different dispensation orders might better resolve the sequence of these regions.
6. The read length of Pyrosequencing is currently limited to 40–50 nucleotides per run in routine analysis.

Acknowledgments

The author is supported by NIH. The author would like to thank Dr. Elahe Elahi for critical reading of this manuscript.

References

1. Ronaghi, M., Uhlen, M., and Nyren, P. (1998) A sequencing method based on real-time pyrophosphate. *Science* **281**, 363–365.

2. Ronaghi, M., Nygren, M., Lundeberg, J., and Nyren, P. (1999) Analyses of secondary structures in DNA by pyrosequencing. *Anal. Biochem.* **267**, 65–71.

3. Nordström, T., Gharizadeh, B., Pourmand, N., Nyrén P., and Ronaghi, M. (2001) Method enabling fast partial sequencing of cDNA clones. *Anal. Biochem.* **15**, 266–271.

4. Garcia, A. C., Ahamdian, A., Gharizadeh, B., Lundeberg, J., Ronaghi, M., and Nyren, P. (2000) Mutation detection by Pyrosequencing: sequencing of exons 5 to 8 of the p53 tumour supressor gene. *Gene* **253**, 249–257.

5. Ahmadian, A., Gharizadeh, B., Gustafsson, A., C., Sterky, F., Nyren, P., Uhlen, M., and Lundeberg, J. (2000) Single-nucleotide polymorphism analysis by Pyrosequencing. *Anal. Biochem.* **280**, 103–110.

6. Alderborn, A., Kristofferson, A., and Hammerling, U. (2000) Determination of single nucleotide polymorphisms by real-time pyrophosphate DNA sequencing. *Genome Res.* **10**, 1249–1258.

7. Nordström, T., Ronaghi, M., Forsberg, L., de Faire, L., Morgenstern, R., and Nyrén P. (2000) Direct analysis of single nucleotide polymorphism on double-stranded DNA by pyrosequencing. *Biotechnol. Appl. Biochem.* **31**, 107–112.

8. Ronaghi, M. (2000) Improved performance of Pyrosequencing using single-stranded DNA-binding protein. *Anal. Biochem.* **286**, 282–288.

9. Ronaghi, M. (2001) Pyrosequencing sheds light on DNA sequencing. *Genome Res.* **11**, 3–11.

10. Ronaghi, M., Karamohamed, S., Pettersson, B., Uhlen, M., and Nyren, P. (1996) Real-time DNA sequencing using detection of pyrophosphate release. *Anal. Biochem.* **242**, 84–89.

14

Homogeneous Allele-Specific PCR in SNP Genotyping

Søren Germer and Russell Higuchi

1. Introduction

1.1. Allele Specificity in Polymerase Chain Reaction (PCR)

Allele-specific (AS) PCR amplification *(1–3)* has been used in combination with gel based detection to genotype-specific polymorphisms. Until recently a major drawback of this method was that it was labor-intensive and without high-throughput instrumentation *(4)*. The single nucleotide polymorphism (SNP) genotyping assay presented here combines AS PCR amplification with kinetic, real-time monitoring *(5–6)*. It is robust, rapid, inexpensive, and allows accurate measurement of allele frequencies in pools of DNA, facilitating large-scale gene mapping.

The allelic specificity of the PCR amplification is conferred by placing the 3' end of one the primers directly over and matching one or the other of the variant nucleotides (*see* **Fig. 1**). Ideally, only completely matched primers are extended and only the matching allele is amplified. In practice, however, there will be more or less amplification of the mismatched allele, as well as nonspecific products such as "primer-dimer" *(7)*. To delay this amplification as much

From: *Methods in Molecular Biology, vol. 212:*
Single Nucleotide Polymorphisms: Methods and Protocols
Edited by: P.-Y. Kwok © Humana Press Inc., Totowa, NJ

A Assay design

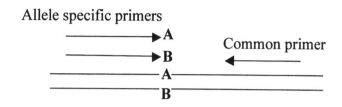

B Amplification

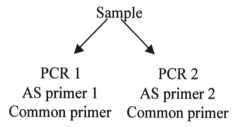

Fig. 1. Allele-specific PCR. (**A**) Allele-specific primers are designed to match, at their 3' ends, the two variants (A/B) at a SNP. (**B**) A sample to be genotyped is divided among two PCRs. One PCR contains one of the two allele-specific primers and the other contains the other allele-specific primer. Both contain the common primer. DNA amplification between an allele-specific primer and the common primer will not occur (or be greatly delayed) if the primer is mismatched to the template.

as possible, we have used the Stoffel fragment of *Taq* DNA polymerase *(5,8)*, which has been shown to enhance discrimination of 3' primer-template mismatches *(9)*. To further minimize the formation of nonspecific products, we use a heat-activatable, "Gold" version of the Stoffel Fragment polymerase to provide a simplified "hotstart" *(10)*. Recently we have further derived a new variant, CEA2 *(11)*, of the Stoffel fragment polymerase. CEA2 also in a "Gold" version, provides improved amplification efficiency without a significant decrease in allelic discrimination (compared to Stoffel; *see* **Note 1**). This has helped us develop hundreds of genotyping

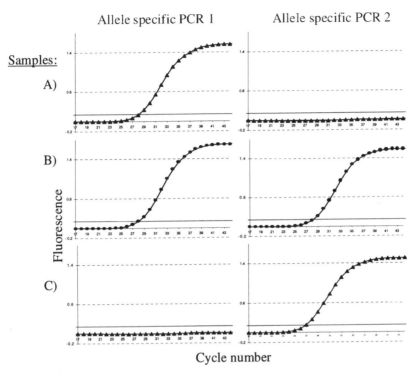

Figure 2. Allele-specific PCR monitored in real-time. Three samples —
(**A**) a homozygous 1 allele, (**B**) a heterozygote, and (**C**) a homozygous 2
allele for a SNP — are split and amplified as diagrammed in **Fig. 1**. The
amplifications are monitored on a cycle-by-cycle basis for increasing
fluorescence owing to the accumulation of dsDNA in the presence of
SYBR green I. The pairs of amplifications for each sample reveal the
genotype. If an allele is not present, no fluorescence increase is detected.

assays for both human and murine SNPs in an efficient and inexpensive manner *(12)*.

1.2. Genotyping of Individual DNA Samples

To genotype an individual DNA sample for a single SNP, an equal
aliquot of the sample is added to two AS PCR reactions, each containing one of the AS primers and the common primer (*see* **Fig. 1**).
PCR reactions containing an AS primer that matches an allele in the

template DNA will amplify normally; PCR reactions containing an AS primer that is mismatched to the alleles in the template will be prevented or delayed. When monitored in real-time (by including SYBR Green I® in the PCR and following fluorescence cycle-by-cycle), PCR growth curves such as we show in **Fig. 2** result. For each amplification a fluorescence threshold near the baseline fluorescence is used to calculate a cycle threshold value (C_t), which is then used to call the genotype of the sample (*see* **Subheading 3.6.**). To validate that a given SNP assay is working, control samples of known genotype should be included in an initial evaluation (*see* **Note 2**).

1.3. Pooling and Allele-Frequency Determination

One approach to high-throughput genotyping of SNPs is to type multiple polymorphisms one individual at a time; for instance, with high-density oligonucleotide hybridization arrays *(13)*. An alternative strategy for typing large numbers of samples and markers is to pool DNA from all the individual samples and then measure the frequency of alleles in the pool one marker at a time. Pooling of DNA samples has been successfully employed with restriction fragment length polymorphisms (RFLPs), microsatellite markers and SNPs *(14–21)*. The present method uses the same primers and amplification conditions as used for genotyping individual DNA samples. Equal aliquots of DNA from a pool are added to each of the two A-S PCR reactions. The PCR reaction containing the A-S primer that matches the allele present in the majority in the pool will amplify earlier (i.e., have a lower C_t) than the PCR reaction containing the AS primer matching the allele present in the minority in the pool (which has a higher C_t). The difference in C_t values between the two (sets of) reactions is proportional to the difference in frequency between the two alleles in the pool (*see* **Fig. 3** and **Subheading 3.6.**). To determine the absolute allele frequency in one pool, it may be necessary to correct for differential amplification efficiency between the two AS primers in a SNP assay (*see* **Note 3**). This is not necessary when the object is to determine the allele frequency differences between two pools, since the resultant bias in the frequency measurement will be the same for both pools.

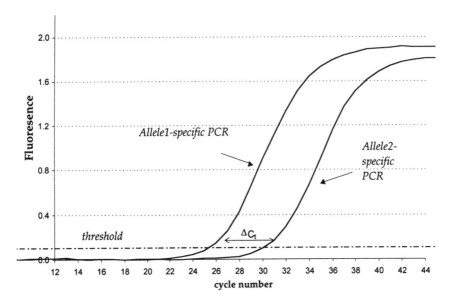

Fig. 3. Allele frequency measurement using kinetic PCR. Equal aliquots of a pool of DNA samples are put into PCRs containing either of the two allele-specific primer sets. The relative allele frequency is determined on the basis of the ΔC_t using the equation: frequency of allele$_1$ = 1/ ($2^{\Delta Ct}$ + 1) (*see* **Subheading 3.6.**).

We have evaluated the accuracy of the kinetic, AS PCR method for many different SNPs and for pools consisting of between 10 and 469 individual DNA samples. The allele frequency determinations are generally accurate for allele frequencies between 2% and 98% (*see* **Note 4**). Statistical analysis of results obtained with different pools and polymorphisms further indicate that for pool sizes up to 1000 samples, the error associated with this genotyping method is generally smaller than the sampling error associated with the number of samples in the pools (for further detail *see* **refs. 6,12,22**). Although care should be taken in the quantitation of the individual DNAs comprising the pools, the routine, small errors commonly seen in DNA quantitation should increasingly cancel out as the number of samples increases. The simplest safeguard against errors arising from the pooling process is to validate the pools by doing, for a few of the SNPs to be screened, genotyping of the individual samples and showing concordance between allele counting and frequency measurement on the pool.

We have tried to provide here standard conditions under which most primer sets will work adequately without optimization. We have developed SNP genotyping assays for more than 500 SNPs. Our first-pass success rate is ~70% under uniform amplification conditions for all the assays.

1.4. Other Homogeneous Detection Methods in PCR

We have also described a single-tube, AS PCR genotyping assay *(5)*. This requires a G-C "clamp" sequence added to the 5' end of one of the two AS primers. Upon completion of the PCR, the T_m of the products is measured *(23)*. By virtue of the G-C clamp, the two allelic products have markedly different T_ms. However, because this analysis is less quantitative than real-time approaches it is not applicable to pooling. For this reason, and because the G-C clamp, depending on sequence context, can inhibit PCR, we have employed the two-tube approach in our work. A number of PCR-based approaches to single-tube, individual genotyping that incorporate homogeneously read, fluorescently labeled, oligonucleotide probes or primers have been developed *(24–30)*. The cost of sequence-specific fluorescent primers or probes for 100s or 1000s of SNPs may be prohibitive, however. Alternatively, generic fluorescent oligonucleotides may be incorporated into the amplicon through tagged primers or probes *(31–36)*. It may be possible to use one or more of these assays to determine allele frequencies in pooled DNA samples.

2. Materials

1. Instrumentation: A kinetic (real-time) thermal cycler. We have used the GeneAmp® 5700 and the ABI Prism® 7700 Sequence Detection Systems (Applied Biosystems). Our collaborators have successfully used the LightCycler™ (Roche Molecular Biochemicals).

2. Polymerase: The Stoffel fragment of *Taq* DNA polymerase is available from Applied Biosystems (AmpliTaq® DNA Polymerase, Stoffel Fragment). We have used this enzyme, as well as a "Gold" version of the Stoffel Fragment, and a "Gold" version of the CEA2 variant (*see* **Subheading 1., Notes 1** and **5**).

3. Primers:
   a. Three oligonucleotide primers for each SNP are required: two AS primers and one common reverse primer. The primer sets can be designed to either of the DNA strands.
   b. Low-cost (approx $30 per assay on average), sephadex desalted (non-HPLC purified) primers made at a low synthesis scale (50 nM) by a commercial oligonucleotide supplier (Operon Technologies, Inc.).
   c. We rely on the primer concentration measurements (by optical density [OD] at 260 nM) supplied by our vendor.

4. SYBR® Green I fluorescent dye (Molecular Probes) at 10,000×. A 20X working stock dilution in dimethyl sulfoxide (DMSO) is made and stored at 4°C. ROX dye (5(6)-Carboxy-X-Rhodamine, Sigma) is added to reduce the contribution of light scatter to the baseline fluorescence measurement so as to minimize well-to-well variation in relative fluorescence increase. ROX should be added in pooling experiments and can be used for genotyping individual samples. A working stock at 200 μM in H$_2$O is made and stored at 4°C.

5. 25 mM MgCl$_2$ (Perkin Elmer).

6. 100 mM dATP, dCTP, dGTP, dTTP, and dUTP (Pharmacia Biotech).

7. 100 mM Tris-HCl, pH 8.0, 500 mM KCl (*see* **Note 6**).

8. 100% DMSO.

9. 80% (w/v) Glycerol.

10. From the commercial stock reagents the following working stocks are made: 2.5 mM each dATP, dCTP, and dGTP; 2.5 mM dTTP; and 2.5 mM dUTP.

11. 1 U/μL Uracil-*N*-glycosylase (Perkin Elmer, AmpErase® UNG). The use of dUTP and UNG is optional, but recommended to control contamination of PCRs by carryover of PCR product from previous work.

12. All PCR reactions are performed in Microamp® Optical PCR tubes (or plates) with Microamp® Optical PCR caps (Applied Biosystems).

3. Methods

3.1. DNA Sample Preparation

1. We have used human and mouse DNA from several sources. Large-scale DNA pools for allele frequency determination were constructed from human DNA samples (100–470 samples per pool) extracted from whole blood *(5,37–38)*; and from mouse genomic DNA samples (~150 samples per pool) prepared from spleen by a salting out method *(39)*, slightly modified. Smaller numbers of DNA samples (~10–20) from mouse and human were pooled for validation studies. Mouse DNA was purchased from Jackson Laboratories and phenol-chloroform extracted from tissue (http://www.jax.org/resources/documents/dnares/), while human DNA samples were prepared from cell cultures with a Puregene® kit (Gentra Systems).

2. For the quantitation of individual DNA samples we have used both OD 260 nm and a DNA specific fluorescent dye, PicoGreen™ (Molecular Probes) following the manufacturer's protocol, and have found both satisfactory. PicoGreen™ detection is more sensitive and specific, but loses sensitivity when DNA is degraded, and it is not appropriate when DNA has been purified with a heating step that dissociates genomic dsDNA.

3.2. Sample Pooling

1. Equal aliquots of individual DNA samples are pooled by pipetting equal volumes of individual DNA samples at equimolar concentrations. For samples of varying concentration, equal amounts (by mass) of each sample are pooled.

2. The quantity needed from each DNA sample depends on the number of samples in the pool(s) and the number of polymorphisms to be tested. With four replicate reactions per allele frequency determination, the quantity can be calculated as 160 ng (i.e., 2×20 ng/rxn $\times$ four replicate reactions) multiplied by the number of polymorphisms, and divided by the number of samples included in the pool. Thus to determine the allele frequency for 100 SNPs in a pool of 1,000 samples, 16 ng of each DNA sample is required.

3.3. Primer Design

1. The two AS primers in a SNP assay are positioned with the base at the 3' terminal end at the SNP position. The common primer is posi-

tioned a short distance from the SNP (*see* **Subheading 3.3.**, **step 5**) and allows amplification with either AS primer. Primers can be designed to either the plus or the minus strand of the sequence.

2. AS primers for different SNP assays can be designed with a similar melting temperature (T_m) *(40)* such that all genotyping assays can be performed under uniform PCR amplification conditions. An anneal/ extend temperature of 58°C has allowed us to design AS primers of appropriate length in both A/T and G/C rich regions. For an anneal/ extend temperature of 58°C, AS primers should have a calculated T_m only slightly higher (59–62°C) to minimize the propensity of the primer to extend a mismatched template. The common primer should be designed with a calculated T_m several degrees above the anneal/ extend temperature (i.e., usually above 63–65°C). This ensures that the common primer will amplify at 58°C, and is likely to tolerate an additional, private polymorphism under the primer sequence.

3. To calculate the T_m we use a modified version of the Nearest Neighbor calculation *(41)* (*see* **Note 7**).

4. The length of the AS primers is determined by the T_m and hence by the sequence surrounding the SNP. Our AS primers are typically in the 16–25 bp range. Even shorter AS primers (14–15 bp) can work. In order to attain a higher T_m, common primers tend to be longer (frequently 20–27 bp) (*see* **Note 8**).

5. Because the PCR conditions are optimized for allelic discrimination, the amplification efficiency of the system is suboptimal. To compensate we design primers that produce a very short amplicon by positioning the common primer <20 bp from the SNP whenever possible (*see* **Note 9**).

6. To evaluate nonspecific primer-dimer amplification potential and excessive hairpin stability in the primers the program Oligo (versions 5.0 or 6.0) can be used. Base pair overlaps >5 bp from the 3' end of the primers should be avoided, especially if there is a high G/C to A/T ratio in the overlapping bases.

3.4. PCR Assembly

1. The number of reactions needed for an experiment is calculated. For individual genotyping the number of PCR reactions is two times the number of individual samples (plus "no template" or other controls). For allele frequency determinations on a pool of DNA samples, four

Table 1
PCR Conditions (100 μL Reaction)

	Stoffel assay		CEA2 assay	
100 m*M* Tris-HCl, pH 8.0	10	μL	10	μL
500 m*M* KCl	8	μL	0	μL
25 m*M* MgCl$_2$	8	μL	12	μL
2.5 m*M* dATP, dGTP, dCTP each	2	μL	2	μL
2.5 m*M* dTTP	1	μL	1	μL
2.5 m*M* dUTP	3	μL	3	μL
Polymerase (12 U/μL)	1	μL	1	μL
UNG (1 U/μL)	2	μL	2	μL
200 μ*M* ROX dye	1	μL	1	μL
20X SYBR Green	1	μL	1	μL
100% DMSO[a]	4	μL	4	μL
80% Glycerol[b]		2.5 μL		2.5 μL
Sterile H$_2$O		50.5 μL		54.5 μL

[a]1% DMSO added with SYBR Green, for a total of 5%.
[b]0.2% Glycerol added with polymerase.

replicate reactions are performed with each AS primer. To determine the absolute allele frequency in one DNA pool, the same
number of replicate reactions is performed on sample (pools) heterozygous for the SNP (*see* **Note 3**). This is not necessary to determine
the difference in allele frequencies between two DNA pools. The
total number of reactions is then two times the number of reactions
for a single DNA pool.

2. For amplifications with Stoffel "Gold" and CEA2 polymerase a basic
   1X mastermix is set up. Multiply the volumes in **Table 1** by the number of reactions (+ 10% to allow for pipetting inaccuracies). 94 μL
   of mastermix is added to each reaction. 2 μL of each of two primers
   (at 10 μ*M*) and 2μL of genomic DNA (at 10 ng/μL) is subsequently
   added to each reaction. The total reaction volume is 100 μL.

3. In practice, a series of submixes are constructed. For each SNP a
   submix is made for each of the two AS primer. The AS primers are
   added to the submix before it is aliquoted into the PCR well. Determine the number of reactions to be performed with each AS primer.
   Multiply that number by 94 μL (+10%); this is the amount of

mastermix to add to the submix. Multiply the same number by 2 μL, and add this volume of one AS primer (at 10 μ*M*) and of the common primer (at 10 μ*M*). The second submix is identical except that the other AS primer is added.

4. For individual genotyping 98 μL of the submix is added to the wells in the optical PCR plate. 2 μL genomic DNA (at 10 ng/μL) from each sample is then added to each of the two wells containing the two submixes. Cover the reactions with optical caps, and skip to **step 7** below.

5. For pooling experiments 2 μL of pooled DNA (at 10 ng/μL), multiplied by the number of replicate reactions (+10%), is added to each of the two submixes containing one of the two AS primers (*see* **Note 10**).

6. For pooling experiments, 100 μL from each submix is added to the wells in the optical PCR plate. For pooling experiments on the ABI 5700 members of pairs of AS PCRs are placed in the same column (as opposed to the same row). This minimizes well-to-well variation. Cover the reactions with optical caps.

7. Place the reaction plates in the thermal cycler, program it with the cycling conditions, and start the PCR amplification.

3.5. Amplification Conditions

1. The PCR cycling conditions are: an initial incubation step of 2 min at 50°C (to allow the UNG-mediated elimination of carryover PCR product contamination *[43]*); an enzyme heat-activation step of 12 min at 95°C; a 45 cycle two-step amplification of 20 s at 95°C and 20 s at 58°C.

2. An on-board dissociation run, from 60°C to 95°C on the ABI 5700, can be performed, and the resulting melting profile for each reaction (or assay) used as quality control for nonspecific amplification (*see* **Note 11**).

3.6. Data Analysis

1. The first step is to determine the C_t (cycle threshold) value for each amplification. The ABI 5700 software (and other similar software) automatically calculates a C_t value based on a user defined arbitrary fluorescence threshold and baseline normalization. The arbitrary fluorescence threshold is set in the "elbow" of the amplification

growth curves; or for amplifications graphed logarithmically, in the middle of the linear phase.

2. Individual genotypes can be determined directly from the C_t values. Amplification of a template DNA with a matching AS primer results in a C_t value between 25 and 32. Amplification of a template with a mismatched AS primer is normally significantly delayed, with a C_t value >35 cycles. The genotype of a given sample can be derived by comparing the C_t values of the two PCR reactions performed on the sample (*see* **Fig. 2**). Subtracting the C_t value of one of the AS PCR reactions from the other results in a ΔC_t value. The ΔC_t value for a given DNA sample should either be a high positive number (>8) for samples homozygous for one allele, a low negative number (<−8) for samples homozygous for the other allele, or close to zero ($1 > \Delta C_t > -1$) for heterozygous samples (*see* **Note 12**). For each particular SNP assay, a range of ΔC_t values from samples of know genotype can be specified and the genotype of unknown samples can be called by comparing the ΔC_t values to that range.

3. For each SNP assay an upper limit for C_t value (e.g., >35 cycles for 20 ng genomic DNA) can be set, such that reactions that generate higher C_t values with both AS primers are automatically counted as amplification failures and excluded from analysis.

4. For the determination of allele frequencies in pooled DNA samples, the ΔC_t values for the two AS PCR reactions for a given pool are determined in **Subheading 3.6., step 2**. For replicate PCR reactions the ΔC_t value is obtained by subtracting the average of the two sets of C_t values. The allele frequency is calculated according to the following formula:

$$\text{Frequency of allele}_1 = 1/(2^{\Delta Ct} + 1),$$

where

$$\Delta C_t = (C_t \text{ of allele}_1 - \text{specific PCR}) - (C_t \text{ of allele}_2 - \text{specific PCR}).$$

The frequency of $\text{allele}_2 = 1 - \text{allele}_1$ frequency.

5. To determine the allele frequency difference between two DNA pools, the allele_1 frequency from one pool is simply subtracted from the allele_1 frequency of the other DNA pool.

6. To determine the absolute allele frequency in a single DNA pool (*see* **Note 3**), the average ΔC_t values obtained from the heterozygote

control samples (or pool) is subtracted from the ΔC_t value obtained from the DNA pool.

7. As a quality control for each SNP assay, the standard deviation of replicate allele frequency determinations should be determined. Allele-frequency determinations with variation above a certain predetermined threshold (e.g., 2 standard deviations) should not be accepted. Too large variability (between allele-frequency determinations) of any particular assay indicates that the AS amplification is not functioning properly. The primers of that particular assay may have to be redesigned.

4. Notes

1. The Stoffel fragment of *Taq* DNA polymerase is available from Applied Biosystems (AmpliTaq® DNA Polymerase, Stoffel Fragment). Stoffel "Gold" and CEA2 "Gold" are not yet commercially available, but can be obtained by contacting the authors at Roche Molecular Systems, Inc.

2. For individual genotyping there are two relevant forms of assay failure: lack of discrimination and amplification failure (or poor amplification). To evaluate the performance of a SNP assay it can be validated on a small set of test samples. When the variant SNP allele is believed to be rare, oligonucleotides matching the amplicon can be synthesized and used as artificial template for validation purposes since the amplicon length is usually <80 bp.

3. To determine the absolute allele frequency for a given pool, it is necessary to correct for the differential amplification efficiency of the two AS primers. This difference in amplification efficiency can be determined by performing the allele frequency determination for a given SNP on a sample (or sample pool) consisting of a 1:1 ratio of each SNP allele. A known heterozygous sample (or a pool of such samples) can be used, and the correction performed as described in **Subheading 3.6.**

4. The two forms of assay failure associated with attempts to determine allele frequency differences between two pools of DNA are not all or none but a matter of degree. They are: 1) failure to discriminate alleles adequately, leading to insensitivity to actual frequency differences between the pools; and 2) excessive assay variability leading to excess type I (false association) errors. For large-scale studies with

1000s of SNP assays, the frequency of the first type of failure may be best estimated by spot-checking. A 20% failure rate means, in essence, that a 10,000 SNP study is actually an 8,000 SNP study. The occurrence of the second type of failure will be known for every SNP by the replicate measurements taken (four for each pool, or eight in all). The SNP-specific variability can be taken into account when assessing the significance of frequency differences at that SNP *(6)*.

5. The Stoffel Fragment can be used under the same conditions listed here for the Stoffel "Gold" reactions, except that the 100 m*M* Tris-HCl should be at pH 8.3 (rather than pH 8.0) and that the heat-activation step should not be included in the cycling conditions.

6. As Stoffel Fragment polymerase is highly salt-sensitive, it is important to ensure that the KCl concentration in the reaction mix is accurate.

7. A public version is available at the Virtual Genome Center (http://alces.med.umn.edu/rawtm.html). Set the parameters to 100 n*M* probe concentration and 100 m*M* salt concentration.

8. Although it has been recommended to avoid primers with a high G/C to A/T ratio, especially near the 3' end of the primer *(42)*, G/C rich primers are preferred over A-T rich primers for the current assay. The amplicons are relatively short and significant amounts of DMSO is added to the reaction such that DNA secondary structure associated with G/C rich template is rarely a problem. Using the Stoffel Fragment polymerase, primers with a high A/T to G/C ratio ($>65\%$) are associated with amplification failure.

9. It may on occasion be necessary to design primers for a longer amplicon in order to retain amplification specificity (e.g., in the presence of sequence homology to regions elsewhere in the genome). But we have generally had poor results with amplicons longer than 90–100 bp.

10. When determining allele frequencies a total of four submixes is made. Two submixes are made for each pool when determining allele frequency differences between two pools, and two for the pooled DNA and two for the heterozygote control(s) when determining absolute allele frequencies using heterozygote controls (*see* **Note 4**).

11. Compared to the temperature peaks of the specific amplicon, primer-dimer amplification usually generates melting curves with a lower temperature peak while amplification of nonspecific homologous genomic DNA may produce longer amplicons with higher temperature melting peaks.

12. A minimum discrimination of >5–6 cycles between samples homozygous for different alleles can be used as a cutoff value for classifying a SNP assay as successful. For some assays there may be a

difference in amplification efficiency between the two AS primer, but that difference is typically minimal compared to the C_t differences between specific and nonspecific allelic amplification and should not affect allele calling in genotyping individual DNA samples.

References

1. Newton, C. R., Graham, A., Heptinstall, L. E., Powell, S. J., Summers, C., Kalsheker, N., et al. (1989) Analysis of any point mutation in DNA: the amplification refractory mutation system (ARMS). *Nucleic Acids Res.* **17**, 2503–2516.
2. Sommer, S. S., Cassady, J. D., Sobell, J. L., and Bottema, C. D. (1989) A novel method for detecting point mutations or polymorphisms and its application to population screening for carriers of phenylketonuria. *Mayo Clin. Proc.* **64**, 1361–1372.
3. Wu, D. Y., Ugozolli, L., Pal, B. K., and Wallace, R. B. (1989). Allele-specific enzymatic amplification of beta-globin genomic DNA for diagnosis of sickle-cell anemia. *Proc. Natl. Acad. Sci. USA* **86**, 2757–2560.
4. Landegren, U., Nilsson, M., and Kwok, P.-Y. (1998) Reading bits of genetic information: methods for single-nucleotide polymorphism analysis. *Genome Res.* **8**, 769–776.
5. Germer, S. and Higuchi, R. (1999) Single-tube genotyping without oligonucleotide probes. *Genome Res.* **9**, 72–78.
6. Germer, S., Holland, M. J., and Higuchi, R. (2000) High-throughput SNP allele-frequency determination in pooled DNA samples by kinetic PCR. *Genome Res.* **10**, 258–266.
7. Chou, Q., Russel, M., Birch, D. E., Raymond, J., and Block, W. (1992) Prevention of pre-PCR mis-priming and primer dimerization improves low-copy-number amplification. *Nucleic Acids Res.* **20**, 1717–1723.
8. Lawyer, F. C., Stoffel, S., Saiki, R. K., Chang, S. Y., Landre, P. A., Abramson, R. D., and Gelfand, D. H. (1993) High-level expression, purification, and enzymatic characterization of full-length *Thermus aquaticus* DNA polymerase and a truncated form deficient in 5' to 3' exonuclease. *PCR Methods Appl.* **2**, 275–287.
9. Tada, M., Omata, M., Kawai, S., Saisho, H., Ohto, M., Saiki, R. K., and Sninsky, J. J. (1993) Detection of *ras* gene mutations in pancreatic juice and peripheral blood of patients with pancreatic adenocarcinoma. *Cancer Res.* **53**, 2472–2474.

10. Birch, D. E. (1996) Simplified hot start PCR. *Nature* **381**, 445–446.

11. Elfstrom, C. M. and Higuchi, R. In preparation.

12. Grupe, A., Germer, S., Usuka, J., Aud, D., Belknap, J. K., Klein, R. F., et al. (2001) In-silico mapping of complex disease-related traits in mice. *Science* **292**, 1915–1918.

13. Wang, D. G., Fan, J.-B., Siao, C.-J., Berno, A., Young, P., Sapolsky, R., et al. (1998) Large-scale identification, mapping, and genotyping of single-nucleotide polymorphisms in the human genome. *Science* **280**, 1077–1082.

14. Arnheim, N., Strange, C., and Erlich, H. (1985) Use of pooled DNA samples to detect linkage disequilibrium of polymorphic restriction fragments and human disease: studies of the HLA class II loci. *Proc. Natl. Acad. Sci. USA* **82**, 6970–6974.

15. Barcellos, L. F., Klitz, W., Field, L. L., Tobias, R., Bowcock, A. M., Wilson, R., et al. (1997) Association mapping of disease loci, by use of pooled DNA genomic screen. *Am. J. Hum. Genet.* **61**, 734–747.

16. Breen, G., Harold, D., Ralston, S., Shaw, D., and St. Clair, D. (2000) Determining SNP allele frequencies in DNA pools. *Biotechniques* **28**(3), 464–466.

17. Kwok, P.-Y., Carlson, C., Yager, T. D., Ankener, W., and Nickerson, D. A. (1994) Comparative analysis of human DNA variations by fluorescence-based sequencing of PCR products. *Genomics* **23**, 138–144.

18. Kwok, P.-Y. (2000) Approaches to allele frequency determination. *Pharmacogenomics* **1**(2), 231–235.

19. Pacek, P., Sajantila, A., and Syvänen, A.-C. (1993) Determination of allele frequencies at loci with length polymorphism by quantitative analysis of DNA amplified from pooled samples. *PCR Methods Applic.* **2**, 313–317.

20. Sasaki, T., Tahira, T., Suzuki, A., Higasa, K., Kukita, Y., Baba, S., and Hayashi, K. (2001) Precise estimation of allele frequencies of single-nucleotide polymorphisms by a quantitative SSCP analysis of pooled DNA. *Am. J. Hum. Genet.* **68**, 214–218.

21. Shaw, S. H., Carrasquillo, M. M., Kashuk, C., Puffenberger, E. G., and Chakravarti, A. (1998) Allele frequency distributions in pooled DNA samples: applications to mapping complex disease genes. *Genome Res.* **8**, 111–123.

22. Chen, J., Higuchi, R., Germer, S., Berkowitz, G., Godbold, J., and Wetmur, J. G. (2001) Kinetic PCR on pooled DNA: a high-through-

put, high-efficiency alternative in genetic epidemiologic studies. *Cancer Epidemiol. Biomark. Prevent.* **11**, 131–136.

23. Ririe, K. M., Rasmussen, R. P., and Wittwer, C. T. (1997) Product differentiation by analysis of DNA melting curves during the polymerase chain reaction. *Anal. Biochem.* **245**, 154–160.

24. Bernard, P. S., Lay, M. J., and Wittwer, C. T. (1998) Integrated amplification and detection of the C677T point mutation in the *methylenetetrahydrofolate reductase* gene by fluorescence resonance energy transfer and probe melting curves. *Anal. Biochem.* **255**, 101–107.

25. Chen, X., Livak, K. J., and Kwok, P.-Y. (1998) A homogeneous, ligase-mediated DNA diagnostic test. *Genome Res.* **8**, 549–556.

26. Fiandaca, M. J., Hyldig-Nielsen, J. J., Gildea, B. D., and Coull, J. M. (2001) Self-reporting PNA/DNA primers for PCR analysis. *Genome Res.* **11**, 609–613.

27. Holland, P. M., Abramson, R. D., Watson, R., and Gelfand, D. H. (1991) Detection of specific polymerase chain reaction product by utilizing the 5'→3' exonuclease activity of *Thermus aquaticus* DNA polymerase. *Proc. Natl. Acad. Sci. USA* **88**, 7276–7280.

28. Kostrikis, L. G., Tyagi, S., Mhlanga, M. M., Ho, D. D., and Kramer, F. R. (1998) Spectral genotyping of human alleles. *Science* **279**, 1228–1229.

29. Tyagi, S. and Kramer, F. R. (1996) Molecular beacons: probes that fluoresce upon hybridization. *Nat. Biotechnol.* **14**, 303–308.

30. Whitcombe, D., Theaker, J., Guy, S. P., Brown, T., and Little, S. (1999) Detection of PCR products using self-probing amplicons and fluorescence. *Nat. Biotechnol.* **17**, 804–807.

31. Beaudet, L., Bedard, J., Breton, B., Mercuri, R. J., and Budarf, M. L. (2001) Homogenous assays for single-nucelotide polymorphism typing using AlphaScreen. *Genome Res.* **11**, 600–608.

32. Jeffreys, A. J., MacLeod, A., Tamaki, K., Neil, D. L., and Monckton, D. G. (1991) Minisatellite repeat coding as a digital approach to DNA typing. *Nature* **354**, 204–209.

33. Myakishev, M. V., Khripin, Y., Hu, S., and Hamer, D. H. (2001) High-throughput SNP genotyping by allele-specific PCR with universal energy-transfer-labelled primers. *Genome Res.* **11**, 163–169.

34. Neilan, B. A., Wilton, A. N., and Jacobs, D. (1997) A universal procedure for primer labelling of amplicons. *Nucleic Acids Res.* **25**, 2938–2939.

35. Whitcombe, D., Brownie, J., Gillard, H. L., McKechnie, D., Theaker, J., Newton, C. R., and Little, S. (1998) A homogeneous fluorescence assay for PCR amplicons: its application to real-time, single-tube genotyping. *Clin. Chem.* **44**(5), 918–923.

36. Winn-Deen, E. S. (1998) Direct fluorescence detection of allele-specific PCR products using novel energy-transfer labeled primers. *Mol. Diagn.* **3**, 217–221.

37. Higuchi, R. (1989) Simple and rapid preparation of samples for PCR, in *PCR Technology: Principles and Applications for DNA Amplification* (Ehrlich, H. A., ed.), M. Stockton Press, New York, NY, pp. 31–38.

38. Helmuth, R., Fildes, N., Blake, E., Luce, M. C., Chimera, J., Madej, R., et al. (1990) HLA-DQα allele and genotype frequencies in various human populations, determined by using enzymatic amplification and oligonucleotide probes. *Am. J. Hum. Genet.* 47, 515–523.

39. Miller, S. A., Dykes, D. D., and Poleskly, H. F. (1988) A simple salting out procedure for extracting DNA from human nucleated cells. *Nucleic Acids Res.* **16**(3), 1215.

40. Wetmur, J. G. (1991) DNA probes: applications of the principles of nucleic acid hybridization. *Crit. Rev. Biochem. Mol. Biol.* **26**, 227–259.

41. Breslauer, K. J., Frank, R., Blocker, H., and Marky, L. A. (1986) Predicting DNA duplex stability from the base sequence. *Proc. Natl. Acad. Sci. USA* **83**(11), 3746–3750.

42. Beasley, E. M., Myers, R. M., Cox, D. R., and Lazzeroni, L. C. (1999) Statistical refinement of primer design parameters, in *PCR Applications: Protocols for Functional Genomics* (Innis, M. A., Gelfand, D. H., and Sninsky, J. J., eds.), Academic Press, San Diego, CA.

43. Longo, M. C., Berninger, M. S., and Hartley, L. L. (1990) Use of uracil DNA glycosylase to control carry-over contamination in polymerase chain reactions. *Gene* **93**(1), 125–128.

15

Oligonucleotide Ligation Assay

Jonas Jarvius, Mats Nilsson, and Ulf Landegren

1. Introduction

1.1. Overview of Current Ligation-Based Single Nucleotide Polymorphism (SNP) Genotyping Approaches

The ability of DNA ligases to join nucleic acids is strongly influenced by mismatches in the ligation substrates *(1–3)*. This mechanism has been exploited in a number of assays where the ability of oligonuleotide probes to be ligated reflects the genotype of the target molecules. This chapter will describe two protocols for solid-phase detection of reaction products in the oligonucleotide ligation assay (OLA), although there are several other detection schemes in use. However, the general considerations of ligase-based sequence distinction are the same, and they will be described in some detail.

The main advantages of the OLA are the reliable discrimination of alleles of SNPs by ligases under a standard set of reaction conditions, and the high specificity of the target detection by pairs of ligation probes, although the latter may not be required in polymerase chain reaction (PCR)-based assays. The reaction can covalently link functions introduced in each of the members of the pairs of oligonucleotide probes to be joined by ligation, allowing the reaction to be

From: *Methods in Molecular Biology, vol. 212:*
Single Nucleotide Polymorphisms: Methods and Protocols
Edited by: P-Y. Kwok © Humana Press Inc., Totowa, NJ

monitored. In the protocols presented herein, target sequences amplified by PCR are interrogated using pairs of ligation probes. One of the members of the probe-pair is equipped with a biotin for capture of the ligation products, and the other with a detectable function. This is a useful combination for solid-phase-based detection formats. However, several alternative formats have been used. One of the probes can be size-coded by different length additions to reflect the identity of the locus, while the allele-specific probe can be color-coded with different fluorescent dyes. Sets of ligation products can then be analyzed using a fluorescence sequencing instrument *(4)*. Alternatively, sequences may be added to one of the probes as a unique sequence tag. Ligation products can then be separated and identified by hybridization to an oligonucleotide microarray, or to fluorescence-labeled microbeads equipped with the complementary tag sequences. The position on the array provides locus identity while the color ratio reflects the genotype *(5,6)*. Ligation products can also be analyzed in homogenous assays by measuring fluorescence resonance energy transfer (FRET) between fluorophores present on the different probes of the probe-pair *(7)*. In a further development of the OLA, the two target-complementary probe sequences may be connected at their distal ends by an extra DNA sequence, creating a circular product upon ligation. These circularized probes become topologically locked onto their target molecules, and they have accordingly been called padlock probes *(8)*. Padlock probes have found utility for *in situ* genotyping *(9)*, and more recently also for PCR-independent SNP analysis using rolling-circle amplification of circularized probes *(10)*. They may also have advantages for parallel analyses of multiple gene sequences without prior target amplification (Banér, unpublished results). Padlock probes will not be discussed further in this chapter, but for more information *see* **refs.** *(10,11)*.

1.2. Practical and Theoretical Considerations for Ligation-Based Assays

Ligation assays can take advantage of the lower hybridization stability of probes mismatched to their target sequences as a mechanism to distinguish similar target sequence variants *(12)*. It is more convenient, however, to use the reduced ability of oligonucleotides to serve as a substrate for enzymatic joining when the ends to be joined are mismatched to their target, as the basis for target sequence distinction *(1,2)*. The *Thermus thermophilus* (*Tth*) ligase can without optimization ligate a correctly base-paired substrate greater than 500-fold more readily than one mismatched in a single nucleotide position *(13)*. In this part of the chapter we will summarize present knowledge about ligases and their mechanism of action. Although numerous ligases have been reported in the literature, we will focus on those most commonly used in ligase-based assays.

1.2.1. Classification of Ligases

DNA ligases have been isolated and characterized from a substantial number of organisms. The DNA ligases can be divided in two major groups accordingly to the cofactor required by the enzyme. Eubacterial ligases require NAD^+ as a cofactor while eukaryotic, archebacterial, and viral ligases require ATP *(14)*. Understanding of the molecular basis for substrate recognition by ligases has increased considerably during recent years. The crystal structure of T7 DNA ligase together with bound ATP revealed striking similarity to the *Chlorella* virus mRNA capping enzyme with a common core structure suggesting mechanistic similarities *(15,16)*. This information served as the basis for more detailed structure-function studies that have clarified the function of the different protein domains and their roles in the ligation process *(17–19)*.

1.2.2. Mechanisms of DNA Ligation

The DNA ligation reaction can be divided in four steps, identical for all characterized DNA ligases. First the enzyme is charged by the covalent attachment of AMP, derived from either NAD^+ or ATP, with the concomitant release of NMN or pyrophosphate, respectively. The AMP molecule is bound to the ε amino group of the lysine residue in a conserved KXDG motif in the ligase. When the enzyme is adenylated it binds to a nicked site in the double-stranded DNA, and it thereafter transfers the AMP molecule to the phosphorylated 5' end at the nick. In the last step the enzyme catalyzes the formation of a phosphodiester bond between the charged 5' end and the 3' OH, releasing AMP.

1.2.3. Basic Considerations for Assay Design

The most widely used enzymes in ligation-based assays are the T4 DNA ligase and the *Tth* DNA ligase. T4 DNA ligase, derived from the bacteriophage T4, is ATP dependent, while *Tth* DNA ligase originates from the eubacterium *Thermus thermophilus* and requires NAD$^+$ as a cofactor. There are some important aspects to consider when setting up a ligation-based assay. The probe-target hybrid must be sufficiently long to accommodate the footprint of the enzyme used. The T7 DNA ligase, which is highly homologous to the T4 DNA ligase, has an asymmetric footprint extending 7–9 nucleotides in the direction of the 5' phosphate and 3–5 nucleotides in the 3' OH direction *(20)*. Further studies have shown that the *Tth* ligase is unable to join oligonucleotides of six nucleotides or less while T4 is able to join six-mers *(21)*. In general a ligation substrate of 20 or more basepairs centered around a nick constitutes as a good substrate for most DNA ligases. Ligation reactions should be performed at a temperature where all oligonucleotides hybridize stably to their complementary targets. It is essential that the ends to be joined have a 5' phosphate and a 3' OH, respectively. Both T4 and *Tth* ligases are more discriminating toward mismatches at the 3' end compared to the 5' end of the ligation junction, reflected in a lower rate of mismatch ligation when the mismatch is placed at the 3' position of the nick *(13,22)*. The *Tth* ligase has been shown to discriminate mispaired bases several nucleotides away from the nick *(21)*. OLA can also be used to distinguish RNA sequence variants *(23)*. The conditions optimal for RNA-templated DNA ligation are quite different from those for DNA sequence analysis, and will not be described in this chapter (optimal conditions for ligation based RNA-sequence analysis, *see* **ref.** *[24]*).

1.2.4. Optimization

In order to minimize mismatch ligation, the minimal amount of ligase required for efficient ligation of matched substrates should be used. T4 and *Tth* ligase differ in cofactor requirement, but also in temperature and pH optima. T4 DNA ligase has a temperature optimum of 37°C, while *Tth* works best between 65–72°C *(22,25–27)*. Divalent cations are required by both enzymes. In general Mg^{2+} is used by both NAD^+ and ATP-dependent ligases. Monovalent cation concentrations can also affect the ligation, and the addition of NaCl at 200 mM enhances the mismatch discrimination of T4 DNA ligase by at least two orders of magnitude *(1,22,28)*. Concentrations of cofactor substantially higher than the K_m for cofactor binding may be helpful to minimize ligation of mismatched substrates. This in analogy to other difficult substrates, such as blunt ends or nicked DNA strands hybridized to RNA. Such reactions are inhibited at high cofactor concentrations owing to premature ATP reloading of the ligase *(24)*. This causes the ligase to dissociate from the substrate after the 5' adenylation step in the ligation reaction, and 5' adenylated nicks are difficult to ligate by adenylated enzymes. The T4 DNA ligase has a K_m for ATP of 14 μM *(29)*, and the *Tth* has a K_m for NAD^+ of 18.5 nM *(27)*. Usually a cofactor concentration of 1 mM is used in ligation assays.

The application of oligonucleotide ligation for SNP genotyping will be exemplified with two assays, published several years ago, that differ in the use of time-resolved fluorescence or an enzyme-linked read-out. The steps of target amplification by PCR, and probe ligation and capture are identical, but thereafter two protocols will be described. All of the reaction steps can be carried out in microtiter wells, and reactions can be handled manually or using a laboratory workstation for increased throughput.

The joining of pairs of oligonucleotides upon hybridization to a target molecule can be conveniently monitored by providing one of the oligonucleotides with a detectable function and the other with a group that can be bound to a solid phase either before or after the ligation reaction (*see* **Fig. 1**). An enzyme-coupled detection reaction

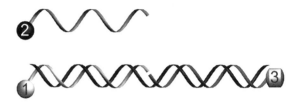

Fig. 1. OLA detects different sequence variants in amplified DNA. The methods uses three different oligonucleotides and a ligase to detect the SNP if interest. One oligonucleotide "3" is used for both sequence variants and is equipped with a function for later capture on a solid support. The two upstream oligonucleotide probes "1 and 2" are differentially labeled and are specific for different target sequence variant. Depending on whether the two alleles are analyzed in separate or the same reaction, probes "1 & 2" can be labeled with the same or with different detectable functions, respectively. In this chapter both time-resolved fluorescence and enzyme-linked readout are described.

can be used and analyzed using standard enzyme-linked immuno-sorbent assay (ELISA)-plate readers, and also time-resolved fluorescence has been applied, with three different ligation probes for each locus, two of them specific for either allelic sequence variant and differentially labelled, and a third in common for the locus of interrest. Using this set-up, alternative sequence variants may be compared in a single reaction, ensuring increased precision of analysis *(30,31)*.

After the ligation reaction, ligation products are captured on a solid support, enabling denaturating washes. Several types of streptavidin-coated solid supports are suitable for the assay, including streptavidin-coated paramagnetic particles (Dynabeads, Dynal AS), streptavidin-coated microplates, or manifold supports *(32)*.

2. Materials

2.1. PCR Amplification

1. Thermal cycler suitable to amplify samples in microtiter plates (e.g., MJ Research or Perkin-Elmer Cetus).
2. 1X PCR buffer: 50 mM Tris-HCl, pH 8.3, 50 mM KCl, 12.5 µg/mL BSA, 1.5 mM MgCl$_2$, 200 µM dNTP and double-distilled water. The MgCl$_2$ concentration can be varied in steps of 0.5 mM to optimize the amplification reaction.

3. Oligonucleotide primers.
4. *Taq* polymerase (Perkin Elmer Cetus).

2.2. Oligonucleotide Ligation Reaction

1. 2X Ligation mix: 18 mM Tris-acetate, pH 7.5, 20 mM magnesium acetate, 100 mM potassium acetate, 400 mM NaCl, 2 mM ATP, 0.4 mU of T4 DNA ligase (Pharmacia), and double-distilled water (*see* **Note 1**).
2. Three labeled oligonucleotides, 600 fmol each per assay.
   a. Time-resolved detection requires one biotinylated and two allele-specific oligonucleotide probes, differentially labeled with europium or terbium chelates.
   b. ELISA-based detection requires one biotinylated and two allele-specific oligonucleotide probes both labeled with digoxigenin (*see* **Note 2**).

2.3. Binding of Ligation Products to a Solid Support

1. Shaking platform (e.g., Perkin Elmer Life Science).
2. Streptavidin-coated paramagnetic particles (Dynabeads, Dynal AS), or streptavidin-coated microtiter plates. Streptavidin-coated microtiter plates can be prepared as follows: regular (Costar, Corning Life Science) microtiter plates are coated with 60 µL, 100 µg/mL streptavidin (e.g., Sigma) in phosphate-buffered saline (PBS) for 2 h at 37°C. The wells are then blocked for at least 30 min at room temperature with 200 µL/well of blocking buffer; 0.5% fat-free dry milk, 100 µg/mL denatured salmon sperm DNA, and 0.02% (w/v) NaN$_3$ in solution B. Microtiter plates can be stored in blocking buffer at 4°C.
3. Solution A: 1 M NaCl, 100 mM Tris-HCl, pH 7.5, 0.1% Triton X-100.
4. Solution B: 150 mM NaCl, 100 mM Tris-HCl, pH 7.5, 0.1% Triton X-100.
5. Denaturing solution: 0.1 M NaOH, 1 M NaCl, 0.1% Triton X-100.

2.4. Time-Resolved Detection of Ligation Products

1. Delfia Plate Reader Research Fluorometer (Perkin Elmer Life Science).
2. Shaking platform (e.g., Perkin Elmer Life Science).
3. Fluorescence enhancement solution for europium or samarium: 0.1 M acetatphthalate, pH 3.2, 15 mM 2-naphtoyl trifluoroacetone, 50 mM tri-N-octylphosphine oxide, and 0.1% Triton X-100 (Perkin Elmer Life Science).

4. Terbium enhancement solution (100 mM 4-(2,4,6-trimethoxy-phenyl)-pyridine-2,6-dicarboxylic acid and 1% cetyltrimethyl-ammonium bromide in 1.1 M NaHCO$_3$ (not commercially available).

5. The two lanthanide labels that we use in this assay, chelates of europium and terbium ions, permit sensitive detection of as little as 0.1 µL of amplification reactions and the two colors are well resolved using a commercially available microplate reader. The synthesis of the labeled probes has been described *(33)*. Probes can also be modified with chelates by reacting amine-modified oligonucleotides, with a reagent commercially available from Perkin Elmer Life Science. The key component of the fluorescence enhancement solution used for detection of terbium ions is not commercially available. The synthesis is described by Hemmilä *(34)*. Samarium chelates can be used in place of terbium chelates. The fluorescence of samarium ions can be recorded in the same enhancement solution used for europium, commercially available from Perkin Elmer Life Science, however samarium ions are detected with approx 10-fold lower sensitivity than europium ions. Other dual-color labels have been developed for use in the oligonucleotide ligation assay, that can be detected in a regular spectrophotometer *(31)*.

2.5. ELISA-Based Detection of Ligation Products

1. ELISA plate reader.
2. Alkaline phosphatase-conjugated anti-digoxigenin antibodies (Boehringer-Mannheim), diluted 1:1000 in solution B before use.
3. Fresh substrate solution for alkaline phosphatase: dissolve a 5 mg tablet of paranitrophenylphosphate (Sigma) in 100 mM diethan-olamin, 0.5 mM MgCl$_2$, pH 9.5.

3. Methods

3.1. Amplification of Target DNA

1. DNA samples to be analyzed for allelic sequence variation at a given position are amplified by PCR, followed by ligase-mediated gene detection.
2. Add to the wells of a microtiter plate 5 µL of genomic DNA at 2 ng/µL in 1X PCR buffer.
3. Add 5 µL of *Taq* polymerase (0.1 U/µL) and primers (2 µM each) in 1X PCR buffer.

4. Subject to 30 temperature cycles, typically 94°C, 55°C, and 72°C, 30 s each.

3.2. Oligonucleotide Ligation Reaction

After amplification the reactions are denatured in the thermocycler. Meanwhile, a ligation mix including three ligation probes and a ligase is added to individual wells of a new microtiter plate (either coated with streptavidin or an uncoated plate if magnetic particles beads are used). After the temperature has reached 37°C, the PCR products are transferred to the new microtiter plate. In the enzyme-linked detection format different allelic forms are detected in separate wells, while TRF detects both allelic variants in the same well. A dual-color ELISA format of the OLA has been described by Tobe et al. *(31)*.

1. For time-resolved decection proceed to **step 2**. For enzyme-linked detection dilute the PCR samples with 60 µL of double-distilled water. Thereafter heat the samples to 96°C for 5 min in order to denature the PCR products.
2. The amplification reactions are heated to 96°C for 5 min to denature the PCR products, and the temperature is then rapidly lowered to 37°C.
3. Immediately add 10 µL of the ligation mix to designated wells in a new microtiter plate.
4. Transfer 10 µL of the denatured PCR samples to the microtiter plate for a total volume of 20 µL.
5. The ligation reactions are incubated for 30 min at room temperature or 15 min at 37°C.

3.3. Binding of Ligation Products to a Solid Support

1. Streptavidin-coated microtiter plates.
   a. Wash the plates twice with solution B to remove unbound streptavidin.
   b. Transfer the ligation reactions to the plate together with 20 µL of solution A.
   c. Incubate 15 min at room temperature or at 37°C.
   d. Wash the plate twice with solution A, and thereafter twice with denaturation solution, followed by a last wash with solution A.

2. Streptavidin-coated paramagnetic particles.
   a. After the ligation step, 20 µL of solution A is added to each ligation reaction.
   b. Two µL of streptavidin-coated paramagnetic particles is added per reaction.
   c. The reactions are incubated on a shaking platform at room temperature for at least 15 min.
   d. The particles are washed twice with solution A by attracting to one side of the wells using a permanent magnet.

3.4. Detection of Ligation Products (see Note 3)

1. ELISA-based readout.
   a. Add 30 µL of antidigoxigenin antibodies in blocking buffer to each well. Incubate for 30 min at 37°C. Therafter wash the wells six times with solution B.
   b. Add 50 µL of substrate solution to each well and incubate at room temperature until the maximal absorbance at 405 nm approaches an optical density of 2.0 (*see* **Note 4**).

2. Time-resolved read-out.
   a. Resuspend the washed particles in the wells of the microtiter plate in 180 µL of europium-fluorescence enhancement solution.
   b. Incubate 10 min on a shaking platform.
   c. Record the europium signals in a Delfia Plate Reader Research Fluorometer.
   d. Next, add 20 µL of a terbium enhancement solution.
   e. Shake 10 min.
   f. Record the terbium signals.

4. Notes

1. The OLA method is capable of detecting and distinguishing all target sequence variants under standard conditions. However the individual experimenter may have to vary the amount of ligase used during OLA in order to find the optimal concentration.
2. One of the oligonucleotides is complementary to a sequence in common for both allelic sequence variant. This oligonucleotide has two modifications, a 5' phosphate for ligation and a 3' biotin used for capture on support. The other two oligonucleotides, each complementary to one or the other of two allelic sequence variants, are

designed to hybridize directly upstream of the first one. For SNP distinction their sequences only differ in one base at the 3' end. The two allele-specific oligonucleotides are further modified at their 5' ends with either digoxigenin for ELISA-based readout or with different fluorescent compounds for the time-resolved readout.

3. If the assay produces a poor signal, individual reaction steps should be investigated. For example, the ligation step of the OLA method can be investigated by labeling one of the oligonucleotides with a radioactive phosphate group in the 5' position using polynucleotide kinase and γ^{32}-ATP. The ligation products analyzed on a 15% denaturing polyacrylamide gel followed by autoradiography. The binding of oligonucleotides to the solid support can be evaluated using doubly labeled oligonucleotides with both biotin and digoxigenin. If a poor signal is reported one may have to improve the streptavidin-coating of the microtiter plates.

4. The intensity of the enzyme-linked detection signal can also be increased by using another substrate for the alkaline phosphatase (e.g., BRL-ELISA detection system *[35]*).

References

1. Landegren, U., Kaiser, R., Sanders, J., and Hood, L. (1988) A ligase-mediated gene detection technique. *Science* **241**, 1077–1080.
2. Alves, A. M. and Carr, F. J. (1988) Dot blot detection of point mutations with adjacently hybridising synthetic oligonucleotide probes. *Nucleic Acids Res.* **16**, 8723.
3. Wu, D. Y. and Wallace, R. B. (1989) The ligation amplification reaction (LAR): amplification of specific DNA sequences using sequential rounds of template-dependent ligation. *Genomics* **4**, 560–569.
4. Grossman, P. D., Bloch, W., Brinson, E., Chang, C. C., Eggerding, F. A., Fung, S., et al. (1994) High-density multiplex detection of nucleic acid sequences: oligonucleotide ligation assay and sequence-coded separation. *Nucleic Acids Res.* **22**, 4527–4534.
5. Gerry, N. P., Witowski, N. E., Day, J., Hammer, R. P., Barany, G., and Barany, F. (1999) Universal DNA microarray method for multiplex detection of low abundance point mutations. *J. Mol. Biol.* **292**, 251–262.
6. Iannone, M. A., Taylor, J. D., Chen, J., Li, M. S., Rivers, P., Slentz-Kesler, K. A., and Weiner, M. P. (2000) Multiplexed single nucleotide polymorphism genotyping by oligonucleotide ligation and flow cytometry. *Cytometry* **39**, 131–140.

7. Chen, X., Livak, K. J., and Kwok, P.-Y. (1998) A homogenous, ligase-mediated DNA diagnostic test for genome analysis. *Genome Res.* **8**, 549–556.

8. Nilsson, M., Malmgren, H., Samiotaki, M., Kwiatkowski, M., Chowdhary, B. P., and Landegren, U. (1994) Padlock probes: Circularizing oligonucleotides for localized DNA detection. *Science* **265**, 2085–2088.

9. Nilsson, M., Krejci, K., Koch, J., Kwiatkowski, M., Gustavsson, P., and Landegren, U. (1997) Padlock probes reveal single-nucleotide differences, parent of origin and *in situ* distribution of centromeric sequences in human chromosomes 13 and 21. *Nat. Genet.* **16**, 252–255.

10. Lizardi, P. M., Huang, X., Zhu, Z., Bray-Ward, P., Thomas, D. C., and Ward, D. C. (1998) Mutation detection and single-molecule counting using isothermal rolling-circle amplification. *Nat. Genet.* **19**, 225–232.

11. Baner, J., Nilsson, M., Isaksson, A., Mendel-Hartvig, M., Antson, D. O., and Landegren, U. (2001) More keys to padlock probes: mechanisms for high-throughput nucleic acid analysis. *Curr. Opin. Biotechnol.* **12**, 11–15.

12. Xu, Y. and Kool, E. T. (1999) High sequence fidelity in a non-enzymatic DNA autoligation reaction. *Nucleic Acids Res.* **27**, 875–881.

13. Luo, J., Bergstrom, D. E., and Barany, F. (1996) Improving the fidelity of *Thermus thermophilus* DNA ligase. *Nucleic Acids Res.* **24**, 3071–3078.

14. Doherty, A. J. and Suh, S. W. (2000) Structural and mechanistic conservation in DNA ligases. *Nucleic Acids Res.* **28**, 4051–4058.

15. Subramanya, H. S., Doherty, A. J., Ashford, S. R., and Wigley, D. B. (1996) Crystal structure of an ATP-dependent DNA ligase from bacteriophage T7. *Cell* **85**, 607–615.

16. Hakansson, K., Doherty, A. J., Shuman, S., and Wigley, D. B. (1997) X-ray crystallography reveals a large conformational change during guanyl transfer by mRNA capping enzymes. *Cell* **89**, 545–553.

17. Sriskanda, V. and Shuman, S. (1998) Chlorella virus DNA ligase: nick recognition and mutational analysis. *Nucleic Acids Res.* **26**, 525–531.

18. Sriskanda, V. and Shuman, S. (1998) Mutational analysis of Chlorella virus DNA ligase: catalytic roles of domain I and motif VI. *Nucleic Acids Res.* **26**, 4618–4625.

19. Doherty, A. J. and Wigley, D. B. (1999) Functional domains of an ATP-dependent DNA ligase. *J. Mol. Biol.* **285**, 63–71.

20. Doherty, A. J. and Dafforn, T. R. (2000) Nick recognition by DNA ligases. *J. Mol. Biol.* **296**, 43–56.
21. Pritchard, C. E. and Southern, E. M. (1997) Effects of base mismatches on joining of short oligodeoxynucleotides by DNA ligases. *Nucleic Acids Res.* **25**, 3403–3407.
22. Wu, D. Y. and Wallace, R. B. (1989) Specificity of the nick-closing activity of bacteriophage T4 DNA ligase. *Gene* **76**, 245–254.
23. Nilsson, M., Barbany, G., Antson, D.-O., Gertow, K., and Landegren, U. (2000) Enhanced detection and distinction of RNA by enzymatic probe ligation. *Nat. Biotechnol.* **18**, 791–793.
24. Nilsson, M., Antson, D.-O., Barbany, G., and Landegren, U. (2001) RNA-templated DNA ligation for transcript analysis. *Nucleic Acids Res.* **29**, 578–581.
25. Weiss, B. and Richardson, C. C. (1967) Enzymatic breakage and joining of deoxyribonucleic acid. 3. An enzyme-adenylate intermediate in the polynucleotide ligase reaction. *J. Biol. Chem.* **242**, 4270–4272.
26. Engler, M. J. and Richardson, C. C. (1982) *DNA Ligases*, Academic Press Inc.
27. Takahashi, M., Yamaguchi, E., and Uchida, T. (1984) Thermophilic DNA ligase. Purification and properties of the enzyme from Thermus thermophilus HB8. *J. Biol. Chem.* **259**, 10,041–10,047.
28. Nilsson, M., Baner, J., Mendel-Hartvig, M., et al. (2002) Making ends meet in genetic analysis using padlock probes. *Hum. Mutat.* **19**, 410–415.
29. Rossi, R., Montecucco, A., Ciarrocchi, G., and Biamonti, G. (1997) Functional characterization of the T4 DNA ligase: a new insight into the mechanism of action. *Nucleic Acids Res.* **25**, 2106–2113.
30. Samiotaki, M., Kwiatkowski, M., Parik, J., and Landegren, U. (1994) Dual-color detection of DNA sequence variants by ligase-mediated analysis. *Genomics* **20**, 238–242.
31. Tobe, V. O., Taylor, S. L., and Nickerson, D. A. (1996) Single-well genotyping of diallelic sequence variations by a two-color ELISA-based oligonucleotide ligation assay. *Nucleic Acids Res.* **24**, 3728–3732.
32. Parik, J., Kwiatkowski, M., Lagerkvist, A., Lagerstrom Fermer, M., Samiotaki, M., Stewart, J., et al. (1993) A manifold support for molecular genetic reactions. *Anal. Biochem.* **211**, 144–150.
33. Kwiatkowski, M., Samiotaki, M., Lamminmaki, U., Mukkala, V. M., and Landegren, U. (1994) Solid-phase synthesis of chelate-labelled

oligonucleotides: application in triple-color ligase-mediated gene analysis. *Nucleic Acids Res.* **22**, 2604–2611.

34. Hemmila, I., Mukkala, V. M., Latva, M., and Kiilholma, P. (1993) Di- and tetracarboxylate derivatives of pyridines, bipyridines and terpyridines as luminogenic reagents for time-resolved fluorometric determination of terbium and dysprosium. *J. Biochem. Biophys. Methods* **26**, 283–290.

35. Nickerson, D. A., Kaiser, R., Lappin, S., Stewart, J., Hood, L., and Landegren, U. (1990) Automated DNA diagnostics using an ELISA-based oligonucleotide ligation assay. *Proc. Natl. Acad. Sci. USA* **87**, 8923–8927.

16

Invader Assay for SNP Genotyping

Victor Lyamichev and Bruce Neri

1. Introduction

In the basic Invader assay, two synthetic oligonucleotides, the invasive and signal probes, anneal in tandem to the target strand to form the overlapping complex shown schematically as primary reaction in **Fig. 1**. The signal probe is designed to have two regions: a target-specific region that is complementary to the target sequence, and a 5' arm or flap that is noncomplementary to both the target and the invasive probe sequences. Structure-specific 5' nucleases *(1)*, known as Cleavase enzymes, recognize this overlapping complex and cleave the signal probe at the site of its overlap with the 3' end of the invasive probe, as shown by the arrow in **Fig. 1** *(2–4)*. This cleavage releases the noncomplementary 5' flap of the signal probe plus one nucleotide of its target-specific region. The cleaved 5' flap serves as a signal for the presence, and enables quantitative analysis, of the specific target in the sample.

The specificity of the Invader assay is determined by the unique ability of Cleavase enzymes to discriminate between substrates that exhibit the overlap between the invasive and signal probes in the complex with the target strand and substrates that lack this overlap. A single nucleotide substitution in the target strand at the position

From: *Methods in Molecular Biology, vol. 212:*
Single Nucleotide Polymorphisms: Methods and Protocols
Edited by: P-Y. Kwok © Humana Press Inc., Totowa, NJ

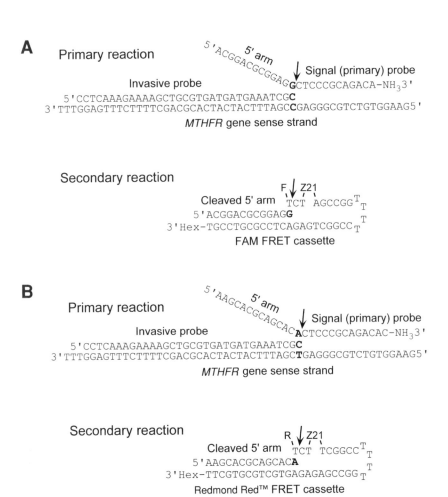

Fig. 1. Invader assay for analysis of the (A) C677 and (B) T677 poly-morphisms in the human methylenetetrahydrofolate reductase (*MTHFR*) gene. The sense strand of the *MTHFR* gene was used for the analysis. The invasive and primary (signal) probes overlap by one nucleotide at the poly-morphic site (shown in bold case). The primary probes have two regions: the target-specific region and the non-complementary 5' flap. The over-lapping complex is specifically cleaved by a structure-specific 5' nuclease at the site of the overlap (shown by arrows) releasing the 5' flap of the primary probe. In the secondary reaction, the cleaved 5' flap from the pri-mary reaction forms an overlapping complex with the FRET cassette. The FRET cassette for the C677 allele includes the FAM (F) dye and Z21 quencher, while the FRET cassette for the T677 allele includes Redmond Red™ (R) and the Z21 quencher. The cleavage sites of the FRET cassettes in the presence of cleaved 5' flaps are shown by arrows.

of overlap eliminates a base pair between the target and the signal probe at this site (shown in bold case in **Fig. 1**) and produces a non-overlapping substrate. The enzymatic cleavage rate of a nonoverlapping substrate is approx 1,000-fold less than that of the overlapping substrate. If only the signal probe anneals to the target (the invasive probe is absent), the enzymatic cleavage rate of this incomplete substrate is more than 300,000-fold less than that of a complete, overlapping substrate *(5)*.

The strong dependence of Cleavase activity on substrate structure makes the Invader assay perfectly suited for detecting small differences, such as single nucleotide polymorphisms (SNPs), in nucleotide sequences of genomes *(6–8)*. To identify a specific nucleotide at a SNP locus, invasive and signal probes are designed to overlap at the site of the SNP. If the signal probe matches the target at this site, the overlapping substrate forms (*see* **Fig. 1**) and the probe is efficiently cleaved, generating the target-specific signal. If the signal probe does not match the target at the SNP site, then the non-overlapping substrate forms and no target-specific signal is generated. For a typical di-allelic SNP, an Invader assay requires two signal probes, one for each of the alternative nucleotides. Only one invasive probe is required for both reactions, however, because the 3' nucleotide of that probe need not basepair with the target and can be any of the four natural nucleotides *(9)*.

To amplify signal, or to generate multiple cleaved 5' flaps per each target molecule, the Invader assay is performed at elevated temperatures corresponding to the melting temperature (T_m) of the signal probe and the reaction mixture contains excess signal probes. Under these conditions, the cleaved signal probe quickly dissociates from the target and is replaced with an uncleaved probe, initiating a new round of cleavage. A typical probe turnover rate (defined as the number of probes cleaved per target per minute) is approx 30 min^{-1}, producing a 10^3–10^4-fold signal amplification in a 2–4 h reaction *(2,5)*.

Combining two Invader reactions, the primary and secondary reactions, in a serial assay, as shown in **Fig. 1**, leads to additional signal amplification. In the serial assay, the signal probes, called primary probes, are cleaved in the primary reaction to produce multiple cleaved 5' flaps. These cleaved 5' flaps participate in the secondary reaction to form another overlapping substrate with a hairpin-like synthetic oligonucleotide, the fluorescence resonance energy transfer (FRET) cassette. The FRET cassette contains a signal and quencher dye pair. Cleavage of the FRET cassette releases the signal dye molecule, which produces a fluorescent signal when it is separated from the quencher. The secondary reaction uses a similar probe turnover mechanism as the primary reaction and each cleaved 5' flap leads to multiple cleaved FRET cassettes. The length and sequence of the 5' flap are designed so that its T_m equals the reaction temperature to ensure the maximal turnover rate. The observed turnover rate for a cleaved 5' flap in the secondary reaction is approx 40 min^{-1} resulting in an additional 10^3–10^4-fold signal amplification. In practice, both the primary and secondary reactions run concurrently in the same reaction tube and generate approx 10^7 free signal dye molecules per target molecule in a 4-h reaction. This level of amplification is sufficient for directly detecting 1,000 copies of a unique genomic DNA sequence per sample if the assay results are read on a standard fluorescence plate reader *(10)*. With a more sensitive instrument, the limit of detection or time to results decreases.

In summary, the serial Invader assay is a homogeneous, isothermal reaction with a fluorescence read-out that can perform direct SNP analysis of samples containing only 10–100 ng of human genomic DNA. As an example, we describe here the design of a serial Invader assay for analyzing the C677T polymorphism in the human methylenetetrahydrofolate reductase (*MTHFR*) gene. The two-dye detection format employed in this assay uses only one well of a microtiter plate for each sample.

2. Materials

1. Sample preparation: Typically, genomic DNA for genetic analysis is obtained from blood samples. We have successfully used the following three DNA extraction kits for isolating DNA from whole blood: the QIAamp Blood kit (Qiagen, Cat. no. 51104), the Puregene kit

(Gentra Systems, Cat. no. D-5500), or the MasterPure kit (Epicentre, Cat. no. MG71100). Any of these kits will produce 3–12 µg of sufficiently purified genomic DNA for the Invader assay. After the final step of extraction, the purified DNA is dissolved in 200 µL of 10 mM Tris-HCl, pH 7.8, and 0.1 mM EDTA to yield a concentration of 15–60 ng/µL and stored at −20°C.

2. Reaction solutions:
   a. Enzyme stock: 200 ng/µL of the structure-specific 5' nuclease Afu FEN1 from *Archaeoglobus fulgidus (2,3)* in 10 mM Tris-HCl, pH 7.8, 50 mM KCl, 100 µg/mL bovine serum albumin (BSA), and 50% glycerol. Store at 4°C.
   b. Reaction buffer (5X): 50 mM MOPS, pH 7.5, and 16% PEG 8000. Store at −20°C.
   c. Magnesium chloride stock (10X): 75 mM MgCl$_2$. Store at −20°C.

3. Microtiter plates and fluorescence microtiter plate reader: The assay described here can be performed in any type of 96-well microtiter plate and results can be read on any fluorescence microtiter plate reader with a sensitivity of 10^9 fluorescein molecules. We recommend using low-profile polypropylene Microtiter Plates 96 (MJ Research, Cat. no. MLL-9601) or equivalent plates and a Cytofluor 4000 fluorescence microtiter plate reader (PE Biosystems, Cat. nos. MIFSOC2TC, MIFS601831). The Invader reactions performed in these microtiter plates can be directly quantitated by a Cytofluor 4000 reader, which eliminates an additional step of transferring the reactions into special read-out plates.

4. Any thermal cycler, dry incubator, or heating block equipped to handle 96-well microtiter plates and capable of maintaining 63°C with precision of ±1°C can be used for the Invader assay described here.

3. Methods

3.1. Design of the Primary and Invasive Probes and FRET Cassette for SNP Genotyping

1. To design an Invader assay for SNP genotyping, the sequence of 40–50 bases on either side of the polymorphic site on the target must be known. Although either the sense or antisense DNA strand can be used, certain features of the probes, such as four or more Gs in a row or sequences that might cause the target-specific region of the primary (signal) probe to form a secondary structure with its 5' flap region, indicate that the opposite target strand should be used instead.

2. Primary probes used in the Invader assay have a 5' flap and a target-specific region. The base at the SNP site on the target DNA determines the base at the 5' end of the target-specific region. In addition, the length of the target-specific region is chosen so that the T_m of the probe-target duplex is approx 63°C. The T_m can be calculated with the Hyther program developed by Peyret and SantaLucia at Wayne State University (http://jsl1.chem.wayne.edu/Hyther/hythermenu. html) or by any similar program using nearest-neighbor parameters for DNA *(11,12)* and including the concentrations of the probe 1 μ*M*. Because the target-specific region of each primary probe will detect only one polymorphic nucleotide at the SNP site, two unique target-specific regions must be designed for a typical di-allelic SNP locus (compare **Fig. 1 A,B**). To complete the primary probe design, the target-specific region is extended at the 5' end with one of the universal 5' flap sequences. These universal 5' flap sequences are independent from the target sequence. As a result, practically any SNP assay can use primary probes designed with different target-specific regions, but the identical two 5' flap sequences. Following these rules, we designed the C-specific (5'-<u>ACG GAC GCG GAG GTC</u> CCC GCA GAC A**C**-NH$_3$-3') and T-specific (5'-<u>AAG CAC GCA GCA C</u>**A**T CCC CGC AGA CAC C- NH$_3$-3') primary probes for the C677T MTHFR polymorphism (*see* **Fig. 1**). The nucleotides complementary to the polymorphic site are shown in bold case and the universal 5' flap sequences are underlined. To minimize background signal, the 3' ends of the primary probes are blocked with an amino group (Glen Research, Sterling, VA).

3. The design of the invasive probe starts with its 3' terminal nucleotide. That nucleotide overlaps with the primary probe's target-specific region at the SNP site and should be noncomplementary to the polymorphic nucleotides at the SNP site, following the order T = C > A > G. Because of this design feature, the identical invasive probe can be used with both primary probes for a particular target. Except for its 3' terminal nucleotide, the invasive probe is complementary to the target. The length of the invasive probe is chosen so that the T_m of the probe-target duplex is approx of 73–78°C or 10–15°C higher than that of the primary probe. Following these rules, the sequence of the invasive probe for the C677T MTHFR assay is 5'-CAA AGA AAA GCT GCG TGA TGA TGA AAT CGC-3' (*see* **Fig. 1**).

4. The two FRET cassettes complementing the 5' flaps of the primary probes complete the design of the Invader assay. Like the 5' flaps, the two FRET cassettes are designed to be universal; the identical FRET cassettes can be used successfully in practically any Invader reaction. The sequences of the FAM (Glen Research) and Redmond Red™ (Epoch Biosciences, Redmond, WA) FRET cassettes developed for the C677T *MTHFR* assay are 5'-FAM-TCT-Z21-AGC CGG TTT TCC GGC TGA GAC TCC GCG TCC GT-Hex-3' and 5'-RedDye-TCT-Z21-TCG GCC TTT TGG CCG AGA GAG TGC TGC GTG CTT-Hex-3', respectively (*see* **Fig. 1**). Both probes use the Z21 group (Eclipse™ Dark Quencher) developed by Epoch Biosciences as a quencher, but dabcyl-dT (Glen Research) can be used instead of Z21. The 3' ends of the FRET probes are blocked with a hexanediol group, Hex (Epoch Biosciences), to reduce background signal from the reaction. As an alternative, the amino group (Glen Research) can also be used here with similar results. Although the assay design described here is for the biplex format, the Invader assay can be easily converted to a monoplex format by using a single FRET probe and two primary probes, each with the same complementing 5' flap sequence. The monoplex format requires two wells of a microtiter plate, rather than one, per sample, however.

3.2. Synthesis and Purification of the Oligonucleotides

1. All oligonucleotides for the Invader assay are synthesized using standard phosphoramidite chemistry and can be ordered from a preferred oligonucleotide vendor. The primary probes and FRET cassettes should be ordered as gel-purified or high-performance liquid chromatography (HPLC)-purified because the products of incomplete synthesis can induce significant non-specific background signal in the assay. The invasive probe can be ordered with a standard desalt/dry-down option.

2. The invasive and primary probes and the FRET cassettes are dissolved in 10 m*M* Tris-HCl, pH 7.8, 0.1 m*M* EDTA and their concentrations are determined by measuring the absorption at 260 nm and using the extinction coefficients 15,400, 7,400, 11,500, and 8,700 A_{260} M^{-1} for A, C, G, and T, respectively. For example, the extinction coefficient of the *MTHFR* invasive probe (*see* **Fig. 1**), which contains

12 As, 5 Cs, 8 Gs, and 5 Ts, is $(12 \times 15,400) + (5 \times 7,400) + (8 \times 11,500) + (5 \times 8,700) = 357,300\ A_{260}\ M^{-1}$ and the concentration of 1 A_{260} solution of this probe is 1 $A_{260}/357,300\ A_{260}\ M^{-1} = 2.8\ \mu M$.

3. The invasive probe is diluted to the final concentration of 1 μM, the primary probes to 20 μM each, and the FRET cassettes to 10 μM each in a 10 mM Tris-HCl, pH 7.8, 0.1 mM EDTA.

3.3. Sample Preparation

The Invader assay usually requires at least 100 ng DNA per assay to analyze a SNP in human genomic DNA using the 96-well microtiter plate format described in this protocol. A 10-μL aliquot of genomic DNA, prepared with a standard kit for DNA extraction from whole blood (*see* **Subheading 2.1.**), is sufficient for the reaction.

3.4. Invader Assay Conditions

1. Deliver 10 μL of each sample of human genomic DNA (15–60 ng/μL) to a separate well of a 96-well microtiter plate (*see* **Note 1**). Add 20 μL of mineral oil (Sigma, Cat. no. M 3516) to prevent evaporation. Use one well as a no-target control by adding 10 μL 10 ng/μL tRNA (Sigma). Denature the DNA by incubating the microtiter plate at 95°C for 5 min.
2. Transfer the microtiter plate to a heating block adjusted to 63 ± 1°C (*see* **Note 2**).
3. Use the stock solutions (**Subheading 2.2.**) to prepare reaction mixture containing 0.1 μM invasive probe, 1 μM of each primary probe, 0.5 μM of each FRET cassette, and 20 ng/μL Cleavase enzyme in 1X reaction buffer.
4. Add 10 μL of the reaction mixture to each DNA sample in the microtiter plate.
5. Mix the reagents in each well by pipeting the solution up and down several times (*see* **Note 3**).
6. Incubate the microtiter plate for 4 h at 63 ± 1°C.

3.5. Data Collection

1. After the reaction is completed, remove the microtiter plate from the heating block and cool it to room temperature. If a fluorescence microtiter plate reader requires a special microtiter plate for reading,

transfer the samples to the reader's microtiter plate. For microtiter plate readers that can use low profile polypropylene Microtiter Plates 96 (MJ Research, Cat. no. MLL-9601) or equivalent plates, the reading can be done directly without sample transfer.

2. To detect the signal from FAM , use 485/20 nm excitation and 530/ 25 nm emission filters. To detect the signal from Redmond Red™, use 560/20 nm excitation and 620/40 nm emission filters. The gains of the reader for the FAM and Redmond Red™ channels should be adjusted to have approximately equal counts for each dye in the no-target control. For the Cytoflor 4000 microtiter plate reader, the typical gain values are 40 and 45 for FAM and Redmond Red™ channels, respectively, and the signals for the no-target control range from 100–200 rfu. The microtiter plate can be read using 10 reads per well. Typically, only 2–3 min are needed to read the results from a 96-well microtiter plate.

3.6. Data Analysis

Table 1 shows the results of C677T *MTHFR* polymorphism analysis in the five samples of human genomic DNA. The data were obtained as described in **Subheading 3.5.** using the Invader assay design shown in **Fig. 1**. The raw FAM and Redmond Red™ signals for each sample were used to determine fold-over-zero (FOZ_C and FOZ_T) values for the C- and T- polymorphisms, respectively, calculated as a ratio of "sample" over "no target" signals. To identify the genotype of each sample, the signal ratio was calculated using FOZ_C and FOZ_T values according to the equation:

$$\frac{|FOZ_C - 1|}{|FOZ_T - 1|}$$

Based on extensive statistical analysis, the following ranges of the signal ratio were selected to identify each of the genotypes:

Signal ratio	Genotype
> 5	C/C
0.3–3.0	C/T
< 0.2	T/T
0.2–0.3 or 3–5	Equivocal (*see* **Note 4**).

Table 1
Analysis of Five Samples of Human Genomic DNA for the
C677T *MTHFR* Polymorphism with the Biplex Invader Assay[a]

	C677 polymorphism		T677 polymorphism			
	FAM signal	FOZ$_T$	Redmond Red™ signal	FOZ$_C$	Signal ratio	Call
No target	111	—	181	—	—	—
Sample 1	163	1.47	312	1.72	0.65	C/T
Sample 2	215	1.94	174	0.96	23.4	C/C
Sample 3	236	2.13	178	0.98	28.2	C/C
Sample 4	210	1.89	404	2.23	0.72	C/T
Sample 5	106	0.95	298	1.65	0.06	T/T

[a]The C677 and T677 alleles were detected by the FAM and Redmond Red™ fluorescence signals, respectively. The fold-over-zero values, FOZ$_C$ and FOZ$_T$, for C677 and T677 alleles, respectively, and the signal ratio values were calculated as described (*see* **Subheading 3.6.**).

Following these criteria, the data shown in **Table 1** identify samples 1 and 4 as heterozygous (C/T) for the C677T polymorphism, samples 2 and 3 as homozygous (C/C) for the C-allele, and sample 5 as homozygous (T/T) for the T-allele (*see* **Note 4–5**).

4. Notes

1. Use disposable aerosol barrier pipet tips to avoid cross-contamination.
2. Use only heating blocks calibrated to ± 1°C.
3. Carefully mix all reagents. The reaction buffer contains 16% PEG, which may not mix well.
4. Theoretically, the signal ratio for heterozygous samples should be close to 1. In practice, this ratio, being relatively constant for any given Invader assay, varies between different assays in a range from 0.3–3. When the signal ratio falls into the equivocal ranges, 0.2–0.3 or 3–5, it usually means that insufficient amount of DNA was used in the assay. It is recommended to measure DNA concentration in the sample and repeat the reaction using larger amount of DNA.
5. The Invader assay described here is highly reliable and usually eliminates the need to perform duplicate assays on the same sample. How-

ever, unexpectedly high signals in both the FAM and Redmond Red™ channels do occur very rarely in some reactions, that would be interpreted as the C/T genotype using the standard analysis. Although the origin of such "outliers" is not completely understood, the results are consistent with the presence of nucleases or DNA-degrading chemicals in a few of the pipet tips or the microtiter plate's wells. We recommend repeat reactions for those samples that produced the unexpectedly high signals in the both channels.

Acknowledgments

We wish to thank Eric Rasmussen for the experimental data and discussions, and Kafryn Lieder for critically reading the manuscript.

References

1. Lyamichev, V., Brow, M. A., and Dahlberg, J. E. (1993) Structure-specific endonucleolytic cleavage of nucleic acids by eubacterial DNA polymerases. *Science* **260**, 778–783.
2. Lyamichev, V., Mast, A. L., Hall, J. G., Prudent, J. R., Kaiser, M. W., Takova, T., et al. (1999) Polymorphism identification and quantitative detection of genomic DNA by invasive cleavage of oligonucleotide probes. *Nat. Biotechnol.* **17**, 292–296.
3. Kaiser, M. W., Lyamicheva, N., Ma, W., Miller, C., Neri, B., Fors, L., and Lyamichev, V. I. (1999) A comparison of eubacterial and archaeal structure-specific 5'-exonucleases. *J. Biol. Chem.* **274**, 21,387–21,394.
4. Kwiatkowski, R. W., Lyamichev, V., de Arruda, M., and Neri, B. (1999) Clinical, genetic, and pharmacogenetic applications of the Invader assay. *Mol. Diagn.* **4**, 353–364.
5. Lyamichev, V. I., Kaiser, M. W., Lyamicheva, N. E., Vologodskii, A. V., Hall, J. G., Ma, W. P., et al. (2000) Experimental and theoretical analysis of the invasive signal amplification reaction. *Biochemistry* **39**, 9523–9532.
6. Ryan, D., Nuccie, B., and Arvan, D. (1999) Non-PCR-dependent detection of the factor V Leiden mutation from genomic DNA using a homogeneous invader microtiter plate assay. *Mol. Diagn.* **4**, 135–144.
7. Mein, C. A., Barratt, B. J., Dunn, M. G., Siegmund, T., Smith, A. N., Esposito, L., et al. (2000) Evaluation of single nucleotide polymor-

phism typing with invader on PCR amplicons and its automation. *Genome Res.* **10**, 330–343.

8. Hessner, M. J., Budish, M. A., and Friedman, K. D. (2000) Genotyping of factor V G1691A (Leiden) without the use of PCR by invasive cleavage of oligonucleotide probes. *Clin. Chem.* **46**, 1051–1056.

9. Lyamichev, V., Brow, M. A., Varvel, V. E., and Dahlberg, J. E. (1999) Comparison of the 5' nuclease activities of *Taq* DNA polymerase and its isolated nuclease domain. *Proc. Natl. Acad. Sci. USA* **96**, 6143–6148.

10. Hall, J. G., Eis, P. S., Law, S. M., Reynaldo, L. P., Prudent, J. R., Marshall, D. J., et al. (2000) Sensitive detection of DNA polymorphisms by the serial invasive signal amplification reaction. *Proc. Natl. Acad. Sci. USA* **97**, 8272–8277.

11. Allawi, H. T. and SantaLucia, J., Jr. (1997) Thermodynamics and NMR of internal G.T mismatches in DNA. *Biochemistry* **36**, 10,581–10,594.

12. SantaLucia, J., Jr. (1998) A unified view of polymer, dumbbell, and oligonucleotide DNA nearest-neighbor thermodynamics. *Proc. Natl. Acad. Sci. USA* **95**, 1460–1465.

17

MALDI-TOF Mass Spectrometry-Based SNP Genotyping

**Niels Storm, Brigitte Darnhofer-Patel,
Dirk van den Boom, and Charles P. Rodi**

1. Introduction

1.1. Overview

Since the invention of dideoxy DNA sequencing *(1)*, the genomes of a number of organisms have been either partially or completely sequenced, including two draft sequences of the human genome *(2,3)*. This offers new opportunities to investigate genetic diversity within and between species. Based on this, an expanding quantity of genetic markers has found widespread use in academic, clinical, and commercial areas.

There are a number of approaches available for investigating genetic variation, including Restriction Fragment Length Polymorphisms (RFLPs; *4*); Amplified Fragment Length Polymorphism (AFLP; *5*); Microsatellites or Short Tandem Repeat Sequences (STRs; *6*); and Single Nucleotide Polymorphisms (SNPs; *7*). SNPs are generally accepted as the most important and valuable marker type owing to their abundance (they are the most common form of

From: *Methods in Molecular Biology, vol. 212:*
Single Nucleotide Polymorphisms: Methods and Protocols
Edited by: P-Y. Kwok © Humana Press Inc., Totowa, NJ

genetic marker), their stability, and their simplicity, which makes assay design and subsequent scoring of these markers simple and straightforward. Application of SNP scoring is found in almost all fields of genomic research and diagnostic routine. They are especially useful in large-scale studies (e.g., pharmacogenomics, linkage studies, and candidate gene association studies) where demand for highly accurate, high-throughput technologies, that can deliver fast, reliable results at low costs has grown dramatically. Since its invention *(8)*, Matrix Assisted Laser Desorption/Ionization Time-of-Flight Mass Spectrometry (MALDI-TOF MS) has been developed and improved to meet these requirements *(9,10)*.

1.2. Principle of MALDI-TOF Mass Spectrometry

SNP scoring by mass spectrometry involves the analysis of intact DNA molecules, thus any decomposition or decay of analyte molecules during the detection process has to be prevented. This is accomplished by embedding the analyte in a crystalline structure of an organic compound or "matrix" *(11)*, which keeps nucleic acids intact during mass spectrometry. Among a number of widely used compounds, 3-hydroxypicolinic acid is the matrix of choice in nucleic acid detection applications *(12,13)*.

Laser bursts of 266 nm or 337nm wavelengths volatilize the analyte-matrix co-crystals. Applied energies range between 1×10^7 and 5×10^7 W/cm^2 and generate a particle cloud (plume), which carries a mixture of charged ions and uncharged molecules. An electric field of approx 30 kV extracts and accelerates the ions. While passing through a field-free drift region of usually 1 m length, the ions are separated by their mass/charge ratio. The particles reach a detector where their time of flight (TOF) is measured. The time of flight is directly proportional to the mass/charge ratio of the analyte. Specifically, for the commonly detected single-charged species: the smaller the mass, the shorter the TOF; the larger the mass, the longer the TOF. Collected data signals are transferred to a computer, which then calculates the respective masses. A depiction of this process is shown in **Fig. 1**.

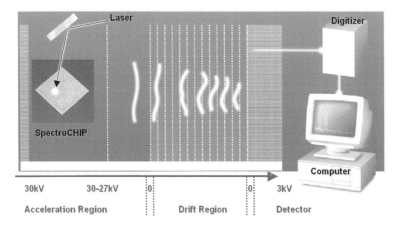

Fig. 1. Principle of MALDI-TOF mass spectrometry. Laser bursts volatize the matrix carrying along with it sample DNA, which is then propelled by an electric field down a flight tube to the detector. The time of flight is proportional to the mass/charge ratio; identically charged molecules with small masses have short times of flight, larger masses have longer times of flight.

1.3. Advantages of MALDI-TOF MS for High-Throughput SNP Scoring

There are a number of features that make MALDI-TOF MS an outstanding platform for scoring SNPs. These include direct detection of the analyte; extremely high analytical accuracy; excellent resolution, which allows good levels of assay multiplexing; simplicity of assay design; and short analysis times. Combined, these features make SNP scoring by MALDI-TOF MS a very high-throughput, inexpensive, and virtually error-free method for investigating genetic diversity. Indeed, mass spectrometric-based methods have been referred to as the "gold standard" for SNP analysis in a recent review on this topic *(14)*.

1.3.1. Simplicity of Assay Design

MALDI-TOF MS has been used for the detection of nucleic acid analytes generated through various methods, including the ligase chain reaction *(15)*, invasive cleavage *(16)*, chemical modification *(17)*, enzymatic nucleic acid fragmentation *(18,19)*, and the polymerase chain reaction (PCR) *(20)*. The simplest and most powerful molecular biological reaction for SNP analysis by mass spectrometry is a primer extension reaction, previously called the PRimer Oligo Base Extension (PROBE) reaction and currently referred to as the MassEXTEND™ assay *(21,22)*. The assay begins with DNA isolation, followed by target amplification (often PCR, though any amplification method is applicable), which provides a first level of stringency; annealing of the extension primer to a site immediately adjacent to the SNP (this provides a second level of stringency); and finally, incorporation of nucleotides across the polymorphic site, which provides a critical third level of stringency. Chain termination is achieved by incorporation of dideoxyribonuclotide triphosphates.

All MassEXTEND assays are designed so that there are at least two dideoxy terminators present and at least one of the alleles is detected by a single base extension. A special case of this is when all four dideoxy terminators are used in the reaction mix, thus yielding only single base extensions for all possible alleles *(23,24)*. Genotypes are called based on the differences between the masses of the terminators corresponding to the possible alleles. For instance, A/T extensions differ by 9 daltons (Da), the smallest mass difference possible; A/C differ by 24 Da; T/G differ by 25 Da; and C/G extensions differ by 40 Da, the largest mass difference. Unfortunately, some of these mass differences are close to the masses of possible ion adducts (Na: 23.0 Da; Mg: 24.3 Da; K: 39.1 Da), which can compromise interpretation of the spectra or require extra processing steps to minimize the occurrence of such adducts. The form of the MassEXTEND assay described here avoids this problem by creating extension products that differ in length in an allele-specific manner, creating mass differences corresponding to the mass of a nucleotide (~300 Da) or more, far in excess of the shifts due to ion adducts. This is illustrated in **Fig. 2** for a C/T polymorphism. The differences in mass between the allele-specific extension products

MassEXTEND Assay

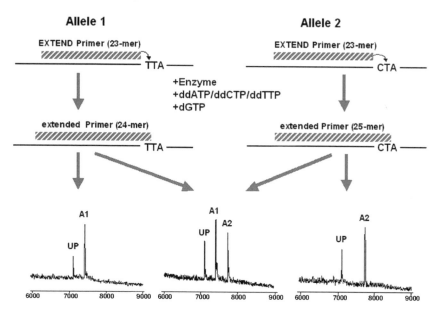

Fig. 2. The MassEXTEND reaction. The key design feature is the use of a terminator mix that yields extension products that differ in length in an allele-specific manner, thus creating mass separations between alleles equal to the mass of a nucleotide. In this example, a normal dG is used along with ddA, ddC, and ddT. For allele 1 (A1), the ddA is incorporated immediately, extending the primer, a 23-mer, to a 24-mer. For allele 2 (A2), the SNP calls for incorporation of the normal dG residue prior to incorporation of a ddA, extending the 23-mer primer to a 25-mer. UP, unextended primer.

(and any unextended MassEXTEND primer) is ~100 times greater than necessary to separate peaks in this portion of the mass spectrum. This makes distinguishing between the different alleles extremely easy.

The next section describes some of the advantages of direct detection of analyte, but it is worth noting here that because MALDI-TOF MS measures molecular mass, an intrinsic property of the allele-specific extension products, there is no need for fluorescent

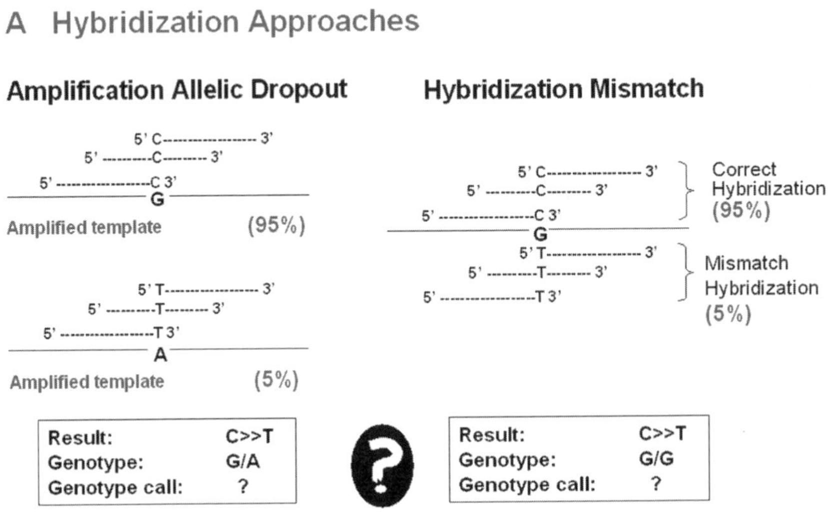

A Hybridization Approaches

Amplification Allelic Dropout   **Hybridization Mismatch**

Fig. 3A. Comparison of hybridization-based reactions to the Mass-EXTEND reaction. (**A**) shows the inability of hybridization across the polymorphic site to distinguish between allelic dropout and hybridization mismatch: in both cases the C-specific signal is much stronger than the T-specific signal. The reporters (in this case, fluoroscence) yield identical signals even though the genotypes are different. In practice, neither of these genotypes can be called with any confidence, since there is no way to discriminate between the two.

dyes (either by themselves or coupled with quenchers), radioactive tracers, or any other form of reporter groups. The use of ordinary oligodeoxyribonucleotide primers without any modification whatsoever, keeps the assay design simple and inexpensive.

1.3.2. Direct Detection of Analyte

The inherent and clear advantage of mass spectrometry is the direct detection of the analyte itself. This is a paradigm shift for molecular biology, because it eliminates all uncertainties caused by indirect detection via labels. This is illustrated in **Fig. 3**, which contrasts hybridization methods using fluorescent reporters, and the MassEXTEND method, which features incorporation of deoxyribonucleotides across the polymorphic site followed by direct detec-

B MassEXTEND

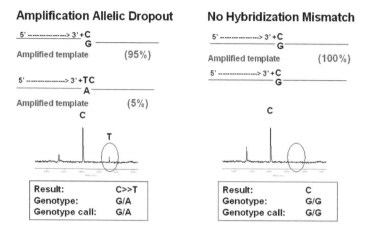

Fig. 3B. **(B)** shows the MassEXTEND assay, which can easily differentiate between the two cases since the annealing does not cover the polymorphism; rather an incorporation of nucleotides across the polymorphic site is needed for any signal to be generated. Because the MALDI-TOF MS is both a separation and detection method, the presence or absence of the T-specific signal at its expected mass position is easily determined by the peak-picking software. Both genotypes are called with confidence.

tion of the analyte by MALDI-TOF MS. **Figure 3A** shows the inability of hybridization across the polymorphic site to distinguish between allelic dropout and hybridization mismatch: in the case of a G/A heterozygote using C-specific and T-specific fluorescent probes, the C-specific signal is much stronger than the T-specific signal due to differential amplification of the G allele over the A allele. The exact same result—that is, the C-specific signal much stronger than the T-specific signal—is seen when the genotype is homozygous G due to the weaker T-specific signal arising from a G:T hybridization mismatch (thermodynamically the most stable mismatch). In practice, neither of these genotypes can be called with any confidence, since there is no way to discriminate between the two. In contrast, the MassEXTEND assay (*see* **Fig. 3B**) can easily differentiate between the two cases since the annealing does not cover the polymorphism; rather an incorporation of nucleotides across the polymorphic site is needed for any signal to be generated.

In the case of allelic dropout, the resulting C-specific signal dominates, but the T-specific signal is easily detected by the peak-picking software, since it is present at its expected mass position. The genotype is correctly called as G/A. In the case of a true homozygous G, no mismatch hybridization is possible, since there is only one hybridizing primer for both alleles and the annealing is a perfect match to sequence adjacent to the polymorphism. Determination of the polymorphism is by incorporation of nucleotides by a high-fidelity polymerase, and in the case of this homozygote, only a C-residue is incorporated. No signal is found in the portion of the spectrum corresponding to the T-specific extension and the correct genotype, homozygous G, is called.

1.3.3. Analytical Accuracy

The analytical accuracy of MALDI-TOF MS is about 0.1–0.01% of the determined mass. The MassEXTEND primers are generally 17–24 nucleotides in length, with extension products rarely greater than 27 nucleotides in length. Using an average mass of 300 Da (dC, 289 Da; dT, 304 Da; dA, 313 Da; dG, 329 Da; avg. = 309 Da), this equates to a typical mass range of 5100–8100 Da. Because each analyte peak in the MassEXTEND assay differs from a neighboring peak by the mass of a nucleotide (again, using 300 Da), this means each peak is between 3.7% and 5.9% different in mass from its neighbor, well in excess of what is needed for the analytical accuracy of MALDI-TOF MS.

This same analysis underscores the advantage of the MassEXTEND assay described here over the single-base extension version described in **Subheading 1.3.1.** (where all four dideoxy terminators are used in the assay). In the single base extension assays, the mass differences between the different allele pairs ranges from 9 Da (dT/dA) to 40 Da (dC/dG) corresponding to differences for an 18-mer of 0.17% to 0.7%; and for a 27-mer of 0.11% to 0.49%. The differences between certain adducts and certain legitimate extensions can be even smaller, presenting a serious challenge to accurate, automated genotype calling.

1.3.4. Multiplexing of Assays

Although the MassEXTEND primers and their extended products typically fall in the range of 17-mers to 27-mers (~5100 Da to ~8100 Da), the masses of these analytes can differ appreciably from one another due to differences in composition. Owing to the simplicity of the assay design, the masses of primers, their allele-specific extensions, and even pausing peaks (*see* **Note 3**) are easily calculated, and multiplexed reactions can be designed and carried out with great confidence. Mass differences for uniplex reactions are typically ~300 Da, but much smaller differences in mass are discernible without running the risk of confusing one peak for another, or confusing an adduct peak for a real peak. Mass differences of 50 Da (corresponding to 0.62% to 1.0% differences in mass within the mass range cited) are routinely scored without sacrificing accuracy. An example of a multiplexed assay for high-throughput genotyping is shown in **Fig. 4**.

These three factors—simplicity of assay design, a large number of masses due to length and composition differences, and excellent analytical accuracy—make significant multiplexing of the Mass–EXTEND reactions possible that easily exceeds that seen with fluorescent-based approaches (limited by available wavelengths) and many gel-based methods (limited by resolving power within the range of primer lengths used).

1.3.5. High-Throughput Genotyping

To make high-throughput genotyping a reality requires simple and fast assay design and processing, plus fast and accurate interpretation and reporting of results. MALDI-TOF MS approaches can encompass all of these features.

1.3.5.1. HIGH-THROUGHPUT ASSAY DESIGN

The simplicity of the MassEXTEND assay design combined with in-house experience that includes literally tens of thousands reac-

Multiplex (5-plex)

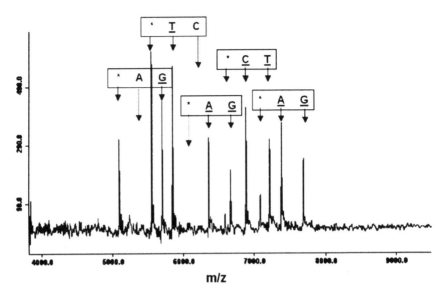

Fig. 4. Multiplexed MassEXTEND assays. Example of a 5-plex analysis of SNPs located on chromosome 22. Peaks specific for a particular SNP are indicated by arrows descending from a common box. In all cases, an asterisk (*) denotes the primer. The genotype is indicated by underlined letter(s) corresponding to the allele(s) that are present. A dotted arrow for a primer means that all of the primer was converted to product. A dotted arrow for an allele simply means that the allele is absent (i.e., the genotype is homozygous for the other allele).

tions has produced design software (SpectroDESIGNER™) that has designed as many as 30,000 assays per computer hour. The designs include both PCR and hME primers, determination of optimal terminator mixes, reaction plate configuration, primer purchase requisitions, and allow multiplexing of reactions.

1.3.5.2. HIGH-THROUGHPUT PROCESSING

Multiplexed homogeneous MassEXTEND (hME) assays are designed as addition-only assays. They are designed as single-tube reactions for both PCR and hME steps. This makes automated assay set up possible and simplifies sample tracking because only a single reaction plate is used up to the transfer of analyte to the SpectroCHIP™ for insertion into the mass spectrometer.

1.3.5.3. HIGH-THROUGHPUT DATA COLLECTION AND GENOTYPE REPORTING

Modern MALDI-TOF MS instruments are capable generating spectra in less than 3.5 s per element examined, including transit time to the next element. Using recently released software (SpectroTYPER RT™) genotypes can be determined using real-time analysis in this same 3.5 s.

1.3.5.4. ACTUAL THROUGHPUT

The combination of factors described in **Subheadings 1.3.5.1.–1.3.5.3.** makes it possible to determine tens of thousands of genotypes in a single day using automated assay set up and a MassARRAY™ system. Management philosophies will of course differ from lab to lab, but if upstream processing of samples is configured in such a way as to make the MALDI-TOF MS the rate-limiting step, then it is possible to determine ~18,240 genotypes in under 4 h using a single MALDI-TOF MS. This is based on the following: 5-plex reactions spotted onto 384-element SpectroCHIPs; 10 SpectroCHIPs per run; an average real-time analysis time of 3.5 s per element; 95% efficiency of multiplexed assays (factors that affect efficiency are similar to those in high-throughput sequencing, e.g., primer quality, template preparation, reagent distribution, etc.).

2. Materials

In the following the materials and protocols for SEQUENOM's hME reaction are listed. The hME assay is especially designed for SNP analysis. It is a single-tube reaction carried out in solution and requires only addition steps throughout the whole procedure. Therefore it is easily compatible with automated liquid handling. The hME reaction is applicable following a genomic PCR amplification procedure and is designed for an automated processing platform specific for SEQUENOM's MassARRAY system.

2.1. Materials Common to Both PCR and hME

The following instruments or components are used to design and process PCR and hME reactions:

1. SpectroDESIGNER software for assay design for PCR and hME reactions (SEQUENOM, Inc., San Diego, CA).
2. Multimek™ 96 Automated 96-channel pipettor (Beckman Coulter, Inc., Fullerton, CA; also available through SEQUENOM as SpectroPREP™). Used with 20 µL tips, also from Beckman Coulter, Inc., Cat. no. 717254.
3. Thermal Cycler: either GeneAmp® PCR System 9700 (Applied Biosystems, Foster City, CA); or PTC-225 DNA Engine Tetrad™ Cycler (MJ Research, Inc., Watertown, MA).
4. Rotator capable of holding microplates (e.g., Fisher Scientific, Pittsburgh, PA; model 346).
5. SpectroPOINT™ (pintool intrument for nanoliter disensing onto SpectroCHIPs) or SpectroJET™ (piezoelectric nanoliter dispenser); both are available through SEQUENOM.
6. MALDI-TOF MS instruments: either Biflex III™ (BRUKER, Bremen, Germany); or Voyager DE™ (Applied Biosystems, Foster City, CA). The instruments are used in connection with SEQUENOM's SpectroTYPER™ and SpectroTYPER RT™ software for data acquisition, automated processing, genotype analysis, and data storage. MS instruments are also available through SEQUENOM with the appropriate software.

2.2. PCR-Specific Materials

1. 384-well microplates (Marsh Biomedical Products, Inc., Rochester, NY, Cat. no. TF-0384).
2. High-performance liquid chromatography (HPLC) grade water.
3. Forward and reverse PCR primers: 30mers, desalted, resuspended in water and stored at −20°C; e.g., from either Integrated DNA Tech-

nologies, Inc. (Coralville, IA); or Operon Technologies, Inc. (Alameda, CA).

4. Ultrapure dNTP set (Amersham Pharmacia Biotech, Inc., Piscataway, NJ; stored at −20°C).
5. HotStarTaq™ DNA Polymerase and buffer (QIAGEN, Inc., Valencia, CA; stored at −20°C).
6. 25 mM MgCl$_2$ (comes with HotStarTaq™ DNA Polymerase).
7. Genomic DNA (2.5 ng/µL; stored at 4°C).

2.3. hME-Specific Materials

1. Autoclaved type I water (resistance >18.2MΩ/cm).
2. Shrimp Alkaline Phosphatase (SEQUENOM; stored at −20°C).
3. MassEXTEND primers (~20-mers, desalted, resuspended in water and stored at −20°C; e.g., from either Integrated DNA Technologies or Operon Technologies).
4. Thermo Sequenase™ DNA Polymerase and buffer (Amersham Pharmacia Biotech; also available through SEQUENOM; stored at −20°C).
5. 10X ddNTP/dNTP Termination mixes (premixed and ready-to-use from SEQUENOM or as single components from Amersham Pharmacia Biotech; stored at −20°C).
6. SpectroCLEAN™ resin for sample desalting prior to mass spectrometry (SEQUENOM; stored at room temperature [RT]).
7. 384-element silicon chip (SpectroCHIP from SEQUENOM, used as platform for MALDI-TOF MS analysis, stored at RT in desiccated environment).

3. Methods

3.1. Assay Design

The assay design is done with SEQUENOM's proprietary software (SpectroDESIGNER). This includes automated PCR primer design as well as hME primer design for singleplex and multiplex reactions. hME primers are designed in a way that, once the primer is hybridized to the PCR template, the 3'-end of the primer is

Table 1
Composition of Uniplex PCR Master Mix

Reagents	Final reaction concentration	One reaction (5 μL)
Water (HPLC grade)	N/A	2.24 μL
10X HotStarTaq PCR buffer containing 15 mM MgCl$_2$ (QIAGEN)	1X buffer containing 2.5 mM MgCl$_2$	0.5 μL
MgCl$_2$ 25 mM (QIAGEN)		0.2 μL
dNTP mix (Amersham Pharmacia), 25 mM each	200 μM each	0.04 μL
Enzyme HotStarTaq DNA Polymerase (5U/μL from QIAGEN)	0.1 U/reaction	0.02 μL
Forward and Reverse PCR Primer mix, 1 μM each primer	200 nM each	1.0 μL
Genomic DNA 2.5 ng/μL	2.5 ng/reaction	1.0 μL
	Sum:	5 μL

immediately adjacent to the polymorphic site of interest. hME primers usually have a length between 17 and 24 bases.

3.2. Assay Protocols

3.2.1. PCR Amplification Protocol

Each PCR reaction has a final volume of 5 μL (*see* **Note 1**). The reactions are setup and performed in 384-well plates. Prepare a PCR master mix without either genomic DNA or PCR primers; the omitted reagent will be added to each reaction individually (*see* **Note 2**). Volumes needed for one reaction are shown in **Table 1**. The PCR program is run using the conditions described in **Table 2**.

Table 2
PCR Program

Temperature	Time	Cycles
95°C	15 min	1
95°C	20 s	
56°C	30 s	45
72°C	1 min	
72°C	3 min	1
4°C	Hold	

3.2.2. Sample Dephosphorylation Protocol

To remove the remaining nonincorporated dNTPs from the PCR reactions, a dephosphorylation step with shrimp alkaline phosphatase enzyme (SAP) is necessary.

1. Prepare SAP working dilution before use: add 0.3 μL of SAP (1 U/μL, SEQUENOM) to 1.7 μL 1:10 diluted Thermo Sequenase reaction buffer.
2. Transfer 2 μL of this SAP working dilution into each 5 μL PCR product, generated in PCR reactions as described in **Subheading 3.2.2.**
3. Incubate for 20 min at 37°C, followed by an inactivation step of 5 min at 85°C.
4. Bring samples to room temperature.

3.2.3. Homogeneous MassEXTEND (hME) Assay Protocol

Prepare the hME cocktail as described in **Table 3**.

1. Add 2 μL of the prepared hME cocktail (as described in **Table 3**) directly to each reaction well after completion of the dephosphorylation step described above.
2. During the thermocycling reaction the hME primers are extended across the polymorphic site (*see* **Note 3**). Use the program described in **Table 4**.

Table 3
hME Cocktail Composition

Reagent	Final conc. in 9 μL reaction	Volume amounts needed per reaction
ddNTPs[a] (Amersham Pharmacia) 10 mM	50 μM each	0.045 μL each
remaining dNTPs[a] (Amersham Pharmacia) 10 mM	50 μM each	0.045 μL each
MassEXTEND primer 100 μM	600 nM	0.054 μL
Thermo Sequenase reaction buffer (Amersham Pharmacia)		0.2 μL
Thermo Sequenase DNA Polymerase (Amersham Pharmacia) 32 U/μL	0.063 U/μL	0.018 μL
Autoclaved water (resistance >18.2MΩ/cm)	N/A	add to 2 μL

[a]ddNTP/dNTP composition is assay specific. The best mix will be chosen automatically by the SpectroDESIGNER software; in case of manual assay design refer to the **Notes** section (#3) for guidelines of optimal termination mix selection.

Table 4
Thermocycling Program for hME

Temperature	Time	Cycles
94°C	2 min	1
94°C	5 s	
52°C	5 s	40 cycles
72°C	5 s	
4°C	Hold	

3.2.4. Sample Cation Cleanup Protocol

After the hME cycling step the samples have to be desalted prior to the mass spectrometry analysis. Therefore 16 μL of water (resistance >18.2MΩ/cm) and 3 mg of SpectroCLEAN resin (SEQUENOM) are added to each reaction. This addition step can be done by using the SpectroPREP 96-channel dispenser (SpectroCLEAN resin and water are added simultaneously) or by using the SpectroPREP only for the water addition, with the SpectroCLEAN resin dumped into the reaction wells by using a 384-dimple plate (SEQUENOM) (*see* **Note 4**).

1. Add 16 μL of water (resistance >18.2MΩ/cm) and 3 mg of SpectroCLEAN resin (SEQUENOM) to each reaction.
2. Place reaction plate on a rotating shaker for 5 min at room temperature.
3. Centrifuge the plate down for 3 min at 1600 rpm (~450g).
4. Transfer about 15 nL of each sample using a nanoliter dispenser (either pintool or piezo electric dispenser, SEQUENOM) onto a 384-element silicon chip preloaded with matrix (3-hydroxypicolinic acid; available as SpectroCHIP from SEQUENOM). The samples dissolve the matrix patch, and, upon solvent evaporation, co-crystallize with the matrix and are ready for MALDI-TOF MS analysis (*see* **Note 5**).

3.3. MALDI-TOF MS Analysis

A linear time-of-flight (TOF) mass spectrometer with delayed extraction is used for the analysis. All spectra are acquired in positive ion mode. Under high vacuum conditions, the matrix crystals are irradiated with nanosecond duration 337-nm laser pulses, leading to formation of a plume of volatilized matrix and analyte as well as charge transfer from matrix ions to analyte molecules. After electric field-induced acceleration in the mass spectrometer source region, the gas-phase ions travel through a ~1 meter field-free region at a velocity inversely proportional to their mass-to-charge. The resulting time-resolved spectrum is translated into a mass spectrum upon calibration. The mass spectra are further processed and analyzed by proprietary software (SpectroTYPER, SEQUENOM) for baseline correction and peak identification. The genotype determination occurs during data acquisition and takes about 3.5 s for each sample, including acquisition and transit time from element to element.

4. Notes

1. As the PCR reaction is performed in only 5 µL, it is important that the TE concentration in the genomic DNA does not inhibit the following reaction. Make sure the genomic DNA does not contain more than 0.25X TE buffer.

2. The Q solution supplied together with the HotStarTaq DNA Polymerase (QIAGEN) should not be used with this protocol. The matrix/ sample crystallization will be disturbed, which decreases the quality of the spectra.

3. It is important to choose the correct ddNTP/dNTP termination mix during the extension reaction of the hME reaction. Occasionally, inappropriate extension products can occur by pausing of the Thermo Sequenase Polymerase after incorporation of one nonterminating nucleotide (i.e., dNTP). This results in a prematurely terminated extension primer, which can confound the analysis if the termination mix is not chosen carefully (e.g., an extension primer elongated with either one ddG or one dA have exactly the same mass and therefore are not distinguishable). The mass difference between a premature termination and a correct termination must be maximized to avoid miscalls. **Table 5** shows the recommended termination mixes for bialleleic SNPs that maximizes the mass difference between the correctly incorporated ddNTP and a correctly incorporated normal dNTP caused by pausing of the polymerase.

4. The desalting step with SpectroCLEAN resin is very crucial for the spectra quality. It is important that the SpectroCLEAN resin particles stay in suspension and do not settle during the 5-min incubation step at RT. Therefore a rotation where the reaction plate gets turned upside down performs best. Increasing either the time or the temperature instead of the rotation is not recommended.

5. In multiplexed reactions the multiplexing occurs at PCR level as well as at hME level. Protocol modifications for multiplex reactions are as follows:

   a. Design. During PCR as well as hME multiplex design, it is important to take primer dimer formation (of each primer involved within one multiplex) into consideration. If you are not

Table 5
Selection of the Optimal Termination Mix[a]

SNP (Biallelic)	Termination mix[b]
A/C	CGT (40 Da)
A/G	ACT (32 Da)
A/T	CGT (25Da)
C/G	ACT (56 Da)
	AGT (24 Da)
C/T	ACG (31 Da)
G/T	ACT (41 Da)
Small ins/del	Dependent on sequence

[a]In the text, genotypes are referred to on the basis of the nucleotide in the template; here they are referred to by the nucleotide incorporated at the +1 site of the extended primer.

[b]Numbers in parentheses are the mass differences between a correct termination and a false termination (i.e., premature termination caused by pausing of the polymerase).

able to use SpectroDESIGNER for your assay design, try to use other programs which check for primer dimers.

b. PCR. Only the PCR primer concentration is reduced in multiplex reactions. The remaining conditions are the same as in singleplex PCR reactions, as shown in **Table 6**. The same PCR program as in singleplex reactions is used.

c. hME. Multiplex reaction conditions are very similar to singleplex reactions. The same reagent compositions are used for all the steps (i.e., dephosphorylation, hME reaction cocktail, desalting). In the hME reaction cocktail, 5 pmol of each primer is added per reaction. Sometimes specific primers give much lower intensity peaks in the mass spectrum. This might be due to concentration errors or due to a different desorption/ionization behavior in the MALDI-TOF MS. Those primers should be adjusted by adding them in a

Table 6
Comparison of Singleplex and Multiplex PCR Setup
(5 µL Total Volume)

Reagent	Singleplex PCR final concentration	Multiplex PCR final concentration
dNTPs	200 µ*M* each	200 µ*M* each
Forward PCR primer	200 n*M*	50 n*M* each
Reverse PCR primer	200 n*M*	50 n*M* each
PCR buffer (QIAGEN)	1X	1X
MgCl₂	2.5 m*M*	2.5 m*M*
HotStarTaq DNA Polymerase 5 U/µL (QIAGEN)	0.1 U/reaction	0.1 U/reaction

higher concentration. The easiest way is to prepare a primer mixture in advance, check it on the MALDI-TOF, adjust it and have it ready-to-use for the actual hME multiplex reactions.

For the hME cycling step an increase from 40–55 cycles improves the reaction. Annealing temperature stays at 52°C.

References

1. Sanger, F., Nicklen, S., and Coulson, A. R. (1977) DNA sequencing with chain terminating inhibitors. *Proc. Natl. Acad. Sci. USA* **12**, 5463–5467.
2. Venter, J. C., Adams, M. D., Myers, E. W., Li, P. W., Mural, R. J., Sutton, G. G., et al. (2001) The sequence of the human genome. *Science* **291**, 1304–1351.
3. Lander, E. S., Linton, L. M., Birren, B., Nusbaum, C., Zody, M. C., Baldwin, J., et al., and International Human Genome Sequencing Consortium (2001) Initial sequencing and analysis of the human genome. *Nature* **409**, 860–921.

4. de Martinville, B., Wyman, A. R., White, R., and Francke, U. (1982) Assignment of first random restriction fragment length polymorphism (RFLP) locus (D14S1) to a region of human chromosome 14. *Am. J. Hum. Genet.* **34**, 216–226.

5. Vos, P., Hogers, R., Bleeker, M., Reijans, M., van de Lee, T., Hornes, M., et al. (1995) AFLP: a new technique for DNA fingerprinting. *Nucleic Acids Res.* **23**, 4407–4414.

6. Taylor, G. R., Noble, J. S., Hall, J. S., Stewart, A. D., and Mueller, R. F. (1989) Hypervariable microsatellite for genetic diagnosis. *Lancet* **2**, 454.

7. Southern, E. M. (2000) Sequence variation in genes and genomic DNA: methods for large-scale analysis. *Ann. Rev. Genom. Hum. Genet.* **1**, 329–360.

8. Karas, M. and Hillenkamp, F. (1988) Laser desorption ionization of proteins with molecular weight masses exceeding 10,000 Daltons. *Anal. Chem.* **60**, 2299–2301.

9. Buetow, K. H., Edmonson, M., MacDonald, R., Clifford, R., Yip, P., Kelley, J., et al. (2001) High-throughput development and characterization of a genomewide collection of gene-based single nucleotide polymorphism markers by chip-based Matrix-assisted laser desorption/ionization time-of-flight mass spectrometry. *Proc. Natl. Acad. Sci. USA* **98**, 581–584.

10. Jurinke, C., van den Boom, D., Cantor, C. R., and Koester, H. (2001) Automated genotyping using the MassARRAY technology. *Methods Mol. Biol.* **170**, 103–116.

11. Karas, M., Glueckmann, M., and Schaefer, J. (2000) Ionization in matrix-assisted laser desorption/ionization: singly charged molecular ions are the lucky survivors. *J. Mass Spectrom.* **35**, 1–12.

12. Bahr, U., Karas, M., and Hillenkamp, F. (1994) Analysis of biopolymers by matrix-assisted laser desorption/ionization (MALDI) mass spectrometry. *Fresenius J. Anal. Chem.* **384**, 783–791.

13. Wu, K. J., Steding, A., and Becker, C. H. (1993) Matrix-assisted laser desorption time-of-flight mass spectrometry of oligonucleotides using 3-hydroxypicolinic acid as an ultraviolet-sensitive matrix. *Rapid Commun. Mass Spectrom.* **7**, 142–146.

14. Weaver, T. (2000) High-throughput SNP discovery and typing for genome-wide genetic analysis. In: *New Technologies for Life Science: A Trends Guide. A Special Issue to Celebrate 25 Years of Trends Publishing*,Wilson, E. et al. eds., Elsevier, Oxford, UK, pp. 36–42.

15. Jurinke, C., van den Boom, D., Jacob, A., Tang, K., Woerl, R., and Koester, H. (1996) Analysis of ligase chain reaction products via matrix-assisted laser desorption/ionization time-of-flight mass spectrometry. *Anal. Biochem.* **237**, 174–181.

16. Griffin, T. J., Hall, J. G., Prudent, J. R., and Smith, L. M. (1999) Direct genetic analysis by matrix-assisted laser desorption/ionization mass spectrometry. *Proc. Natl. Acad. Sci. USA* **96**, 6301–6306.

17. Sauer, S., Lechner, D., Berlin, K., Lehrach, H., Escary, J.-L., Fox, N., and Gut, I. G. (2000) A novel procedure for efficient genotyping of single nucleotide polymorphisms. *Nucleic Acids Res.* **28**, E13–e-13

18. von Wintzingerode, F., Böcker, S., Schlötelburg, C., Chiu, N. H. L., Storm, N., Jurinke, C., et al. (2002) Base-specific fragmentation of amplified 16S rRNA genes analyzed by mass spectrometry analysis: a novel tool for rapid bacterial identification. *Proc. Natl. Acad. USA* **99**, 7039–7044.

19. Hartmer, R., Clemens, J., Storm, N., Böcker, S., Hillenkamp, F., van den Boom, D., and Jurinke, C. (2001) New high throughput approach for sequence analysis via base-specific RNA cleavage reaction. Poster at ASMS Conference 2001, Chicago, IL.

20. Siegert, C. W., Jacob, A., and Koester, H. (1996) Matrix-assisted laser desorption/time of flight mass spectrometry for detection of polymerase chain reaction products containing 7-deazapurine moieties. *Anal. Biochem.* **243**, 55–65.

21. Braun, A., Little, D. P., and Koester, H. (1997) Detection of CFTR gene mutations by using primer oligo base extension and mass spectrometry. *Clin. Chem.* **43**, 1151–1158.

22. Little, D. P., Braun, A., Darnhofer-Demar, B., and Koester, H. (1997) Identification of apolipoprotein E polymorphisms using temperature cycled primer oligo base extension and mass spectrometry. *Eur. J. Clin. Chem.* **35**, 545–548.

23. Haff, L. H. and Smirnov, I. P. (1997) Single nucleotide polymorphism identification assays using a thermostable DNA polymerase and delayed extraction MALDI-TOF mass spectrometry. *Genome Res.* **7**, 378–388.

24. Li, J., Buttler, J. M., Tan, J., Lin, H., Royer, S., Ohler, L., et al. (1999) Single nucleotide polymorphism determination using primer extension and time of flight mass spectrometry. *Electrophoresis* **20**, 1258–1265.

Index

From: *Methods in Molecular Biology, vol. 212:*
Single Nucleotide Polymorphisms: Methods and Protocols
Edited by: P-Y. Kwok © Humana Press Inc., Totowa, NJ